© Copyright 2009

Written by Sally A Jones
Edited by Annalisa Jones

Published by GUINEA PIG EDUCATION

ISBN: 978-0-9561150-9-6

2 Cobs Way
New Haw
Surrey
KT15 3AF

This book is a fun book designed for children to use at home, or to reinforce work completed in school. It is aimed at KS2, KS3 levels 3-5, and children of 9-13 depending on ability.

The handwritten calculations, backed up by the delightful cartoon characters, are part of the books appeal. It is designed to help children grasp basic mathematical concepts in a fun way.

The author has trialled the books with the children she teaches.

It is best to record your answers in an exercise book, but you can fill in the book if you wish.

Contents

Meet the Gang
From Mathsland High School

Patsy Perfect's shopaholic dad won the lottery, but she is a sensible girl with an eye for a bargain.

Doubtful Darren likes to get loads of doughnuts, but he has to divide them between his desperate dogs.

Melissa Mean has more 'moggies' than she can manage.

FUNNY FRANK FRACTION

Matt Multiple and Felicity Factor and her ferrets

Petite Penny Prime and Pony

Naughty Natalie with Nuisance

Long Lily and Lois

DARING DONNA DECIMAL
RIDES THE COURSE

Mighty Measures

Angry Andrew argues over area ...

Loud Lenny Long Division has learnt not to panic...

Work out with Thomas

Ask Amit Angle

Katy Kilogram keeps cooking

Learn that 1,000, 000 has 7 digits.

```
1
1    0
1    0    0
1    0    0    0
1    0    0    0    0
1    0    0    0    0    0
1    0    0    0    0    0    0
```

One million has 7 digits One million 1, 000, 000		Ten thousand 10, 000 One thousand 1, 000 One hundred 100 One ten 10 One unit 1
	> greater than 8 > 5 < less than 7 < 9	
One hundred thousand has 6 digits. 100, 000		

Million	Hundred Thousands	Tens of thousands	Thousands	Hundreds	Tens	Units
1	4	5	3	6	0	1

(no tens ... put 0)

One million, four hundred and fifty three thousand six hundred . . . . . and one.

Now write these numbers:

a) One million, five hundred and twenty one thousand two hundred and three

b) Nine hundred and sixty five thousand, four hundred and twenty eight.

c) Seventy thousand and forty six

d) Five thousand one hundred and thirty eight

e) Six million two hundred and eighty seven thousand, four hundred and thirty six

f) Six hundred and five thousand two hundred and forty five

g) One million and thirty five thousand four hundred and one.

h) Which is more   367,929   or   376,214
   Which is less   431,265   or   430,264

i) Arrange these numbers in order (smallest first).

a) 1,182   789   8,645   846   149   6,032

b) 1,400   10,450   1,236   100,452   100,000   7,060

Work through pages 9 – 20 to see if you know basic maths facts.

$8 + 9 =$ $13 + 2 =$ $5 + 15 =$

$7 + 8 =$ $14 + 5 =$ $11 + 9 =$

$5 + 13 =$ $10 + 6 =$ $12 + 6 =$

$12 + 4 =$ $11 + 7 =$ $7 + 5 =$

$6 + 8 =$ $17 + 2 =$ $17 + 3 =$

$3 + 14 =$ $9 + 10 =$ $14 + 6 =$

$5 + 15 =$ $6 + 7 =$

$4 + \boxed{} = 12$ $13 + \boxed{} = 19$ $6 + \boxed{} = 11$

$12 + \boxed{} = 17$ $1 + \boxed{} = 19$ $14 + \boxed{} = 20$

$7 + \boxed{} = 14$ $18 + \boxed{} = 19$ $9 + \boxed{} = 15$

$10 + \boxed{} = 16$ $9 + \boxed{} = 18$ $13 + \boxed{} = 18$

$9 + \boxed{} = 14$ $14 + 1 = \boxed{}$ $\boxed{} + 2 = 12$

$\boxed{} + 8 = 18$ $\boxed{} + 6 = 13$ $\boxed{} + 5 = 14$

Now turn these add sums to take away sums. For example, if
$8 + 9 = 17$ and $9 + 8 = 17$ then $17 - 9 = 8$ and $17 - 8 = 9$

26 + □ = 39   28 + 31 = □

15 + □ = 40   19 + 19 = □

36 + □ = 54   75 + 25 = □

27 + 14 = □   29 + 49 = □

19 + 18 = □   47 + 28 = □

58 + □ = 90   61 + 19 = □

46 + □ = 100   84 + 16 = □

67 + □ = 99   78 + 19 = □

□ + 25 = 50   35 + 27 = □

30 + 25 = □   29 + □ = 46

□ + 50 = 75   37 + □ = 94

38 + □ = 78   26 + 26 = □

61 + □ = 81   □ + 14 = 38

74 + □ = 93   □ + 47 = 85

27 + □ = 42   □ + 21 = 65

39 + 26 = □   37 + 37 = □

Four Rules

+ = add, plus or total

− = subtract, takeaway or minus

× = multiply, product or times

÷ = divide or share

< = less than

> = greater than

Use BIDMAS to work out brackets

Brackets (indices) ÷, ×, +, −

30 − (8 × 3)
↓
Always work out the brackets first

Make sure you remember the Four Rules.

+ add plus more	− take away subtract minus less
× times multiply product of	÷ divide share

$7 \times 8 = 56$ $7 + 5 = 12$

$8 \times 7 = 56$ $5 + 7 = 12$

$56 \div 7 = 8$ $12 - 7 = 5$

$56 \div 8 = 7$ $12 - 5 = 7$

Learn the Hard Table:

16 to learn

6 x 6	7 x 6	8 x 6	9 x 6
6 x 7	7 x 7	8 x 7	9 x 7
6 x 8	7 x 8	8 x 8	9 x 8
6 x 9	7 x 9	8 x 9	9 x 9

Multiples are like tables. Complete these:

2

3, 6, 9, 12, 15, 18, 21, 24

4

5

6

7, 14,   21,

8

9

10

Learn the Basics

Factors

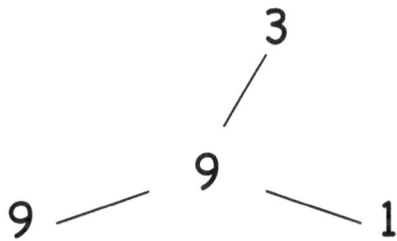 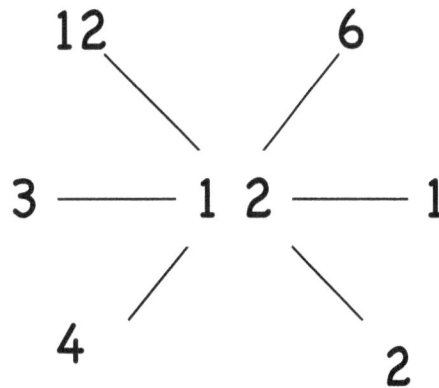

3 x 4 = 12

2 x 6 = 12

1 x 12 = 12

Now find the factors of 21, 15, 24, 30, 60

Prime numbers have **2 factors** – themselves and 1

2,   3,   5,   7,   11,   13,   17,   19,   23,   29

Square numbers

1 (1×1),   4 (2×2),   9 (3×3),   16 (4×4),   25,   36

Cubed numbers

1 (1×1×1),   8 (2×2×2),   27,   64,   125

Triangular numbers

1 (+2),   3 (+3),   6 (+4), 10 (+5), 15 (+6), 21

0.5 = 1/2   1/4 =

0.25 = 1/4   1/2 =

0.75 = 3/4   3/4 =

0.1 = 1/10   1/10 =

0.01 = 1/100   1/100 =

1/5,   2/10,   20/100,   20%,   0.2

2/5,   4/10,   40/100,   40%,   0.4

3/5,   6/10,   60/100,   60%,   0.6

4/5,   8/10,   80/100,   80%,   0.8

5/5,   10/10, 100/100,   100%   1.0

2/10,   20/100,   20%,   0.2

Can you complete the pattern?

3/10 =   5/10 =   6/10 =   8/10 =   9/10 =

$$\frac{3}{4} \xleftarrow{\text{numerator}} \qquad \frac{3}{4} \text{ of } 40 = (40 \div 4) \times 3$$

$$\frac{3}{4} \xleftarrow{\text{denominator}} \qquad \qquad \qquad 10 \times 3$$

LEARN THIS.

What is 30% off a book costing £20.00?

We say:

$$10\% = £2$$
$$30\% = £2 \times 3 = £6.00$$
$$£20.00 - £6.00 = £14.00$$

To find 10% ÷ by 10

$$20 \div 10 = £2$$

$$30\% = 10\% \quad 10\% \quad 10\%$$

$$£2 \times 3$$

THIS METHOD IS HARDER BUT TRY TO UNDERSTAND.

$$\frac{30}{100} \times \frac{20}{1} = £6.00$$

$$£20.00 - £6.00 = £14.00$$

$$30 \div 100 \times 20 =$$

$$0.3 \times 20 =$$

To find a percentage

Joel scored 20 out of 50 in a test.

$$\frac{20}{50} = \frac{40}{100} = 40\% \qquad \text{or} \qquad 20 \div 50 \times 100$$

Sophie scored 30 out of 50. What percentage did she score?

To turn a decimal into a fraction

a) $0.7 = \dfrac{7}{10}$　　b) $0.75 = \dfrac{75}{100}$　　c) $0.625 = \dfrac{625}{1000}$

Convert the Metric Measures

```
1000g = 1kg
500g = ½ kg
250g = ¼ kg
750g = ¾ kg
```

These follow the same pattern as above:

```
1000 ml = 1 litre
500 ml =
250 ml =
750 ml =
```

```
1000 grams = 1 kg
500 g =
250 g =
750 g =
```

but

100 cm = 1 metre	10 mm = 1 cm
50 cm = ½ m	1000 mm = 1m
25 cm = ¼ m	75 cm = ¾ m

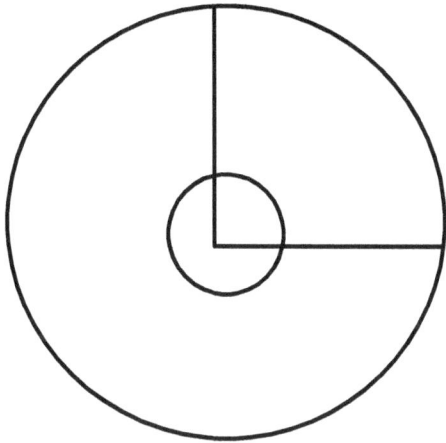

A circle measures 360°

A right angle measures 90°

¾ of a circle measures ....

A straight line measures 180°

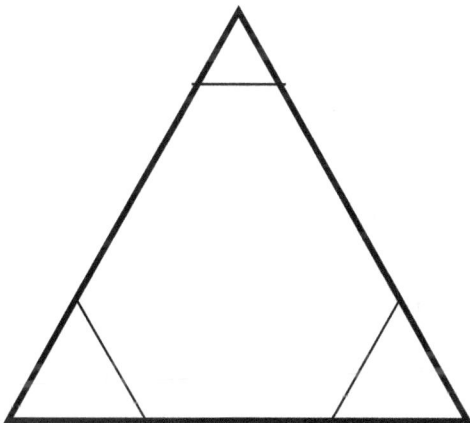

The angles of a triangle add up to 180°

Find out these facts. Can you draw these triangles?

What is an equilateral triangle?

What is an isosceles triangle?

What is a scalene triangle?

Anita adds anything ambitiously.

She uses the column method.

Anita has 3 alligators that she feeds. How much do they eat?

900 + 500 + 200 = 1600g
60 + 40 + 60 = 160g
Total = 1760g

960g
540g
260g +
_____

g

_____

1000 g = 1 kg
÷ by 1000

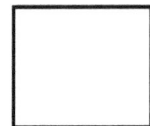 kg

---

Susie and her baby 'gators'

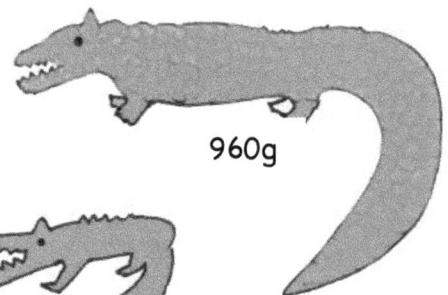

540g

960g

260g

Anita goes to the pet shop. How much does she spend?

She buys:

3 collars   £4.56
             £3.28 +
             £2.64
             _____

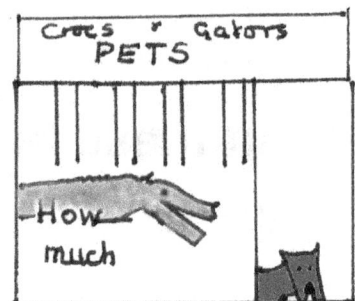

3 bowls   £3.50 + £4.20 + £2.70 =

Look at the sign

Complete + or -

568	6692
369 +	4861 +
————	————
542	6478
79 -	4245 -
————	————
300	400
164 -	237 -
————	————
5368	5000
1979 +	1896 -
————	————
4000	7359
3217 -	1864 +
————	————

Do you know this?

53	42
42 +	53 +
————	————

95	95
42 -	53 -
————	————

$5 \times 6 =$ $5\overline{)30}$

$6 \times 5 =$ $6\overline{)30}$

Work through this page. Look at the signs. If you get the subtractions wrong, work through pages 23-25. Come back and do the corrections.

21

- Subtract
- Minus
- Take away

What is the Difference?

Sam
Subtracts

borrow 1000	borrow 100	borrow 10	take it from the 9	
³4̸	¹⁰1̸	¹⁸9̸	¹2	Regroup
2	3	9	4	
1	7	9	8	

Check on calculator

What is the difference between

4037 and 2864?

Subtract 2391 from 3654?

What is 5238 minus 3400?

4862
3000 -
_____

Take 2.6 from 3.2

³4̸ ⁹0̸ ⁹0̸ ⁹0̸ ¹0
-      3 9 5 4
_____

Take
2.4 from 4.83

Subtract 235
From these numbers

- Using mental arithmetic
- Check on the calculator
- Do a subtraction sum

| 503 | 200 | 30 | 5 | **Answer** = 268 |

203

173

168

Check

$$\require{cancel}\begin{array}{ccc} {}^{4}\cancel{5} & {}^{9}\cancel{0} & {}^{1}3 \\ 2 & 3 & 5 \\ \hline 2 & 6 & 8 \end{array} -$$

| 716 | 823 | 645 |
| - 235 | - 235 | - 235 |

Find the difference

547 and 362   and   362 and 243

362 to 370 = 8
370 to 400 = 30
400 to 547 = 147

147
+ 30   That
8   makes ...
185

$$\begin{array}{ccc} {}^{4}\cancel{5} & {}^{1}4 & 7 \\ - 3 & 6 & 2 \\ \hline 1 & 8 & 5 \end{array}$$

23

"Now make sure you really can subtract," says Sam

85	67	24
− 35	− 23	− 18

70	61	631
− 25	− 18	− 358

531	511	644
− 259	− 294	− 357

934	548	542
− 869	− 374	− 357

5726	5810	5327
− 4487	− 996	− 4245

200 − 153 =   600 − 423 =   7000 − 568 =

1000 − 834 =   4020 − 3153 =   6003 − 428 =

Mad Matt Multiple

Multiplies of a number are just like times table.

So here go the multiples of four …

4, 8, 12, 16, 20, 24, 28, 32, 36, 40, 44, 48, 52, 56, 60, 64

and on and on forever …

Write the Multiples of 6 …

6, 12, ..., ..., ..., ..., ..., ..., ..., ..., ..., ...

And his faithful friend Felicity Factor finds factors
multiply together to make other whole numbers.

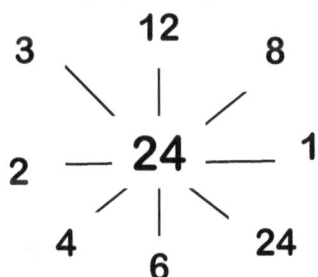

```
   3      12      8                              3, 4        1, 12
    \     |     /        You can write             \         /
 2 — 24 — 1           them in pairs                  12
    /     |     \        3 x 4 = 12                    |
   4      6      24                                  2, 6
```

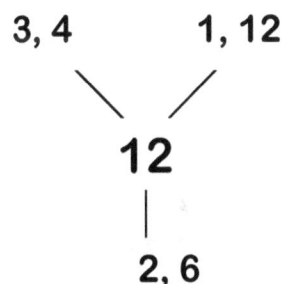

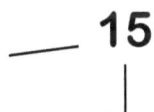

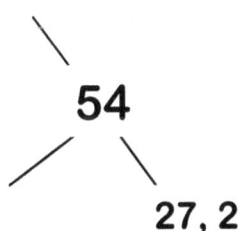

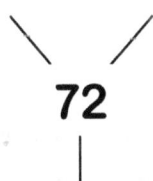

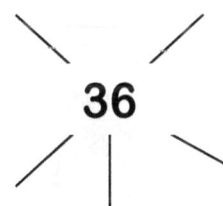

```
        54          72         21        36
```

```
  — 15
    |
```

```
        27, 2
```

The highest
common factor
is the highest
one they both
share = 6.

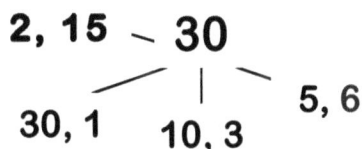

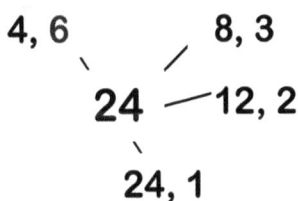

```
2, 15 ~ 30                    4, 6      8, 3
  /    |   \                    \       /
30, 1  10, 3  5, 6             24 — 12, 2
                                  \
                                 24, 1
```

25

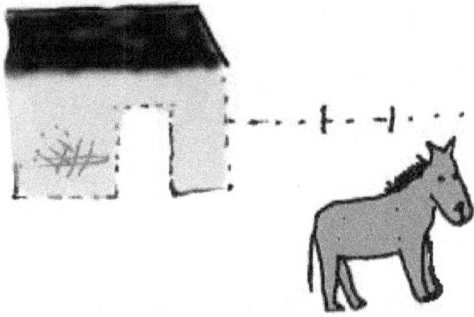

PETITE PENELOPE PRIME

Write all your numbers from 1-100 in the grid.

	2	3							
11									
21									
31									

1 has only 1 factor so it's not a prime number.

2 is the only even prime number.

Prime numbers have two factors: themselves and one.

Ring the prime numbers. Use this method.

Cross out multiples of
2
3
5
7

1 x 2   1 x 5
1 x 3   1 x 7

26

Number Sequence

Naughty Natalie knows she must notice the difference between the numbers in a sequence...

23 28 33 38 43 ☐

What is the rule?

1   4   9   16   25 ☐☐

What is the rule?

1   3   6   10   15 ☐☐

What is the rule?

70 63 56 49 35 ☐☐

What is the rule?

95 89 83 77 ☐☐

What is the rule?

7   8   10   13   17 ☐☐

What is the rule?

8   16 24 32 40 ☐☐

What is the rule?

8   18 28 38 48 ☐☐

What is the rule?

Morgan multiplication
must multiply

10 x 4 =   1 x 5 =   6 x 3 =   8 x 4 =

1 x 2 =   9 x 4 =   7 x 4 =   2 x 5 =

8 x 2 =   7 x 3 =   7 x 2 =   2 x 2 =

6 x 4 =   9 x 2 =   8 x 3 =   10 x 3 =

3 x 3 =   6 x 5 =   10 x 2 =   5 x 4 =

3 x 2 =   3 x 5 =   9 x 3 =   8 x 5 =   4 x 2 =   2 x 4 =

7 x 5 =   6 x 2 =   9 x 5 =   3 x 4 =   5 x 3 =   2 x 9 =

4 x 4 =   10 x 5 =   1 x 4 =   1 x 3 =   2 x 3 =   1 x 7 =

3 x 6 =   2 x 7 =   7 x 7 =   4 x 7 =   1 x 6 =   2 x 6 =

4 x 6 =   3 x 7 =   5 x 7 =   6 x 7 =   1 x 8 =   4 x 8 =

5 x 8 =   2 x2 =   3 x 8 =   1 x 9 =   3 x 9 =   4 x 9 =

| 5 x 8 = 40 so 40 ÷ 5 = 8 | Write each one down in this way |
| 8 x 5 = 40 so 40 ÷ 8 = 5 | in your exercise book. |

28

1 x 11 =

12 x 11 =

Multiply
9 by 6

2 x 11 =

Find the product
of 8 and 6.

8 x 11 =

4 x 11 =

6 x 11 =

9 x 11 =

Find the product
of 12 and 8

3 x 11 =

10 x 11 =

12
x 4
___

12 x 12 =

12 x 10 =

12 x 6 =

2 x 12 =

12 x 5 =

12
x 3
___

12
x 8
___

12 x 11 =

12
x 9
___

12 x 7 =

1 x 12 =

Add 10 then Add 2

12 24 36

Test your skills

1)   Write your prime numbers to thirty

2)   Write your square numbers to 100

3)   Write the first six cubed numbers

4)   Write the first ten triangular numbers

5)   Write down the multiples of

   - six

   - eight

   - seven

   - nine

   - twelve

6)   Write the factors of these numbers:

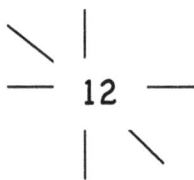
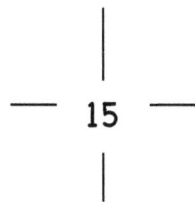
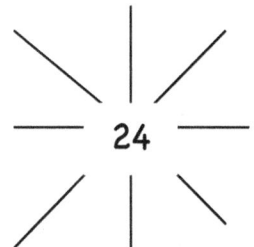
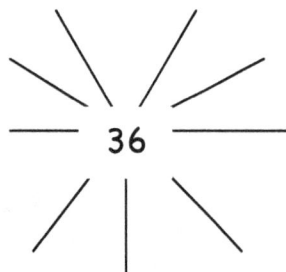
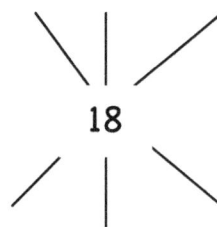

6 x 6 =   7 x 6 =   8 x 6 =   9 x 6 =

6 x 7 =   7 x 7 =   8 x 7 =   9 x 7 =

6 x 8 =   7 x 8 =   8 x 8 =   9 x 8 =

6 x 9 =   7 x 9 =   8 x 9 =   9 x 9 =

Square Tables

1 x 1 =   2 x 2 =   3 x 3 =   4 x 4 =

5 x 5 =   6 x 6 =   7 x 7 =   8 x 8 =

9 x 9 =   10 x 10 =

Count on

6, 12, 18 ....

7, 14, 21, 28 ....

15, 30, 45 ....

50, 100, 150 ....

250, 500, 750, 1000 ....

60, 120, 180 ....

90, 180, 270, 360 ....

$^1/_2$, 1, 1 ½, 2, 2 ½, 3 ....

¼, ½, ¾, 1, 1 ¼, 1 ½ ....

31

Use column method:

Use grid method to check if you wish.

1

$$\begin{array}{r} 23 \\ \times \quad 6 \\ \hline \mathbf{138} \end{array}$$

6 x 3
6 x 20

$$\begin{array}{r} 29 \\ \times \quad 7 \\ \hline \end{array}$$

$$\begin{array}{r} 28 \\ \times \quad 4 \\ \hline \end{array}$$

$$\begin{array}{r} 426 \\ \times \quad 8 \\ \hline \end{array}$$

$$\begin{array}{r} 33 \\ \times \quad 8 \\ \hline \end{array}$$

$$\begin{array}{r} 325 \\ \times \quad 7 \\ \hline \end{array}$$

$$\begin{array}{r} 4372 \\ \times \quad 9 \\ \hline \end{array}$$

Use mental arithmetic

$25 \times 7 =$ $20 \times 7 = 140$ Total $\begin{array}{r} 25 \\ \times \ 7 \end{array}$
              $7 \times 5 = 35$ 175

$36 \times 4 =$

$82 \times 6 =$

$31 \times 5 =$

32

1
2
3
4
5
6
7
8
9
10
11
12
13
14
15
16
17
18
19
20
21
22
23
24
25
26
27
28
29
30
31
32
33
34
35
36
37
38
39
40
40
41
42
43
44
45
46
47
48
49
50
51
52
53
54
55
56
57
58
59
60
61
62
63

Colour in the 7 times table.

Doubtful Darren divides 63 delicious doughnuts between his 7 desperate dogs.

33

63 doughnuts divided between 7 dogs is ...

$63 \div 7 =$

$$\frac{63}{7}$$

$7\overline{)63}$

Count on 7

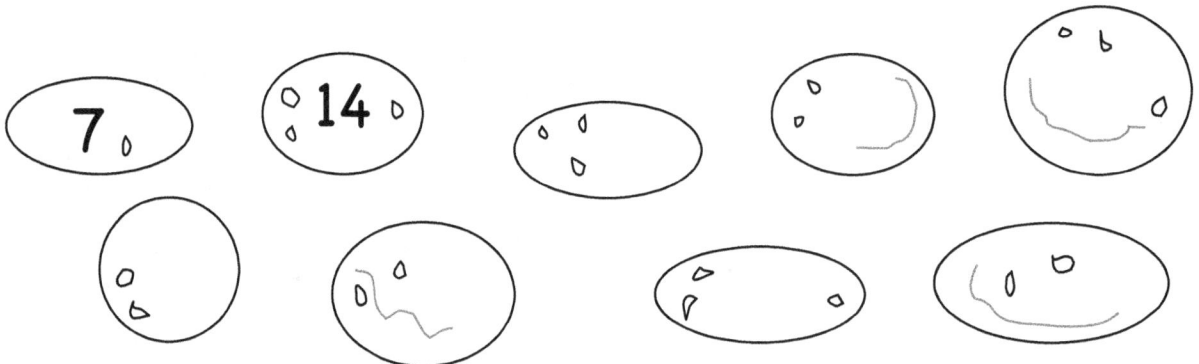

7 14

How many 7s in 63?

because

$7 \times \square = 63$

$\square \times \square = 63$

"Don't give dogs doughnuts! It is BAD FOR THEM. They must eat healthily."

VET

I'm overweight...

34

Share these bags of doughnuts between the 7 dogs

Do this for each bag:

$14 \div 7 = 2$

14

28

35

21

42

49

56

133

119

70

77

63

98

105

Use this method:

$$\begin{array}{r} 19 \\ 7\overline{)13\,^63} \end{array}$$

Use a bus shelter

$1 \times 7 = 7$

remainder

● ● ● ● ● ●

$7 \times \square = 63$

Make up some more examples.

Share 56 toy mice between 8 cute cats.

Share 54 nuts between 9 good guinea pigs.

You many check these divisions using your own method if you wish.

There were actually 69 doughnuts in the sack. If doubtful Darren divides 69 doughnuts between his 7 desperate dogs, how many will they have each? How many are left?

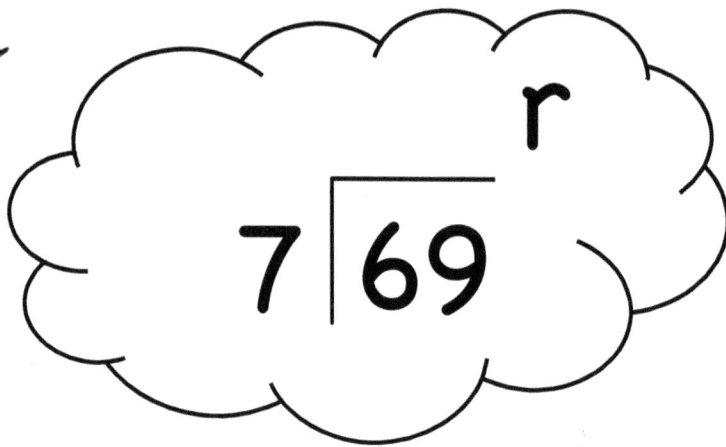

$$7\overline{)69}^{\ r}$$

Because:

9 x 7 =

69 – 63 =

A stray dog joins the family.

How many doughnuts will
the desperate dogs get now?

$8\overline{)69}$

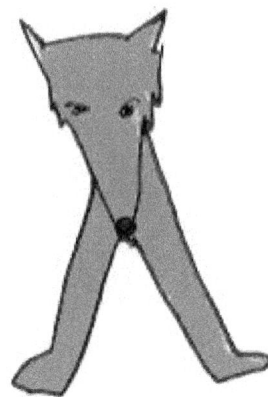

Now try these:

$7\overline{)50}$ $9\overline{)84}$ $9\overline{)64}$ $5\overline{)39}$

$4\overline{)79}$ $6\overline{)59}$ $7\overline{)490}$ $8\overline{)104}$

$6\overline{)156}$ $9\overline{)146}$ $8\overline{)1036}$ $5\overline{)732}$

Use this **traditional method.** Do you know other methods for working out long multiplication? Try working out the answers your way. Which way works best?

Long Lily works out on Long Multiplication

	45		16		36		125
×	32	×	28	×	24	×	42
	90						
	1350		0				
	1440						
	1						

WORK
OUT
HERE

2 x 5 carry the 10
2 x 4 10's + one 10
lay an egg
3 x 5 units, carry 10
3 x 4 10's + the 10
add

Lay an egg
or
Add a zero

38

Work out Here.
Long Division

$$35 \overline{)770}$$
$$\frac{70}{7}$$

$$\begin{array}{r} 35 \\ \times \ 2 \\ \hline 70 \end{array}$$

Write out the
35 times table.

Do these…

$$18 \overline{)432}$$

Write out the
18 times table.

$$20 \overline{)640}$$

Write out the
20 times table.

$$14 \overline{)336}$$

Write out the
14 times table.

Long Lily works on Long Division

$$16 \overline{)240}$$

Write out the
16 times table.

For example:

$$
\begin{array}{r}
.25 \\
15 \overline{)375} \\
30 \\
\hline
75 \\
75 \\
\hline
0
\end{array}
$$

How many sets of 15 in 3 = 0

How many sets of 15 in 37 = 2 (because 15 x 2 = 30)

Take away 37 – 30 = 7   Put 2 in answer

Bring Down 5   Put 5 in answer

How many 15s in 75 = 5 (because 15 x 5 = 75)

Lennie Long Division has learnt not to PANIC

Use long multiplication to find the missing number e.g. 43 x 19 =
16 x 45 =

```
        43              45
   19 │ 81         16 │ 72
```

Check on a calculator.

Writing out the multiples of 19 and 16 can help you work out long division.

19	38	57	76	95	114	133	153	171	190
16	32								

```
           4 3
   19 │ ⁷8 ¹1 7
        7 6
       _____
        0 57
       _____
          57
       _____

   16 │ 72
```

```
    19          19
  x  4        x  3
  ____        ____
   76          57
    3

    43          45
  x 19        x 16
  ____        ____
  ____        ____
```

A D.V.D player costs £266 if Lennie
Long division pays £19.00 a week.
How many weeks will it take him?

A video player costs £204. Lennie pays £17 a
week. How many weeks will it take him?

Lennie buys a widescreen TV at
£1408 over 22 weeks. How much
does he pay each week?

He buys a play station. He
pays £12 a week for 15 weeks.
How much does it cost?

Write the multiples

17   34

22   44   66

15   30   45   60

```
                48
    17 ⟋⁷8̶ ¹1 6
        68
       ___
       136
```

```
            49
    14 ⌐6 ▢ 6
```

| 17 | 34 | 51 | 68 | 85 | 102 | 119 | 136 | 153 | 170 |

| 14 | 28 |

```
    48            49
  × 17          × 14
  ____          ____

  ____          ____

  ____          ____
```

Don't Panic

Think

```
        28
    14 ▢ 9 2
       28
       112
       112

        36
    15 ▢ 4 0
```

```
    14            36
  × 28          × 15
  ____          ____

  ____          ____

  ____          ____
```

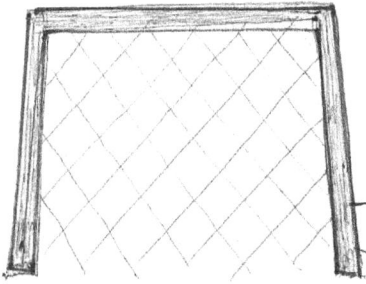

LEARN NAMES

$$\frac{5}{11}$$

5 ← NUMERATOR

11 ← DENOMINATOR

FRACTION WORDS

choc bar

for after match

$\frac{3}{4}$ = 0.

eaten left

$\frac{1}{2}$ = 0.5

eaten left

Funny Frank Fraction

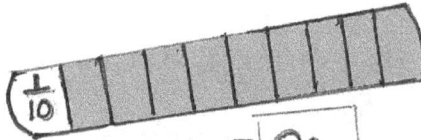

= 0.

finds football fun

but **FRACTIONS**

FEARFUL

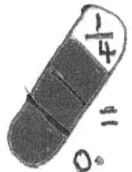

= 0.

☐ eaten
☐ left

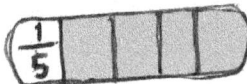

$\frac{1}{5} = \frac{2}{10} = \frac{}{100}$ = 0.

43

THE FOOTBALL FEAST

Colour and count the fractions

Fraction of black rats

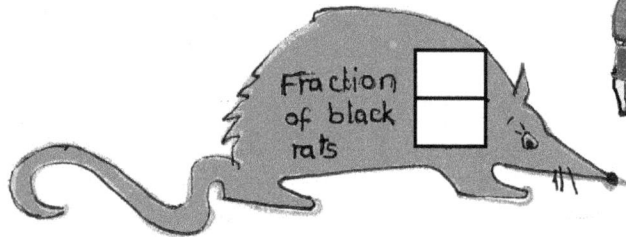

Ate 10
of 16

Ate 6

Ate 12

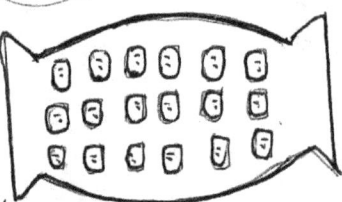

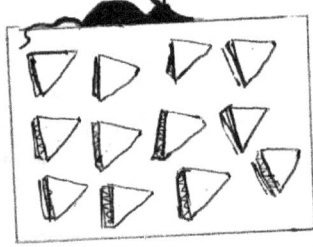

Drank 5

Ate 3

$$\frac{6}{12} \frac{\div 6}{\div 6} = \frac{1}{2}$$

**Learn to cancel down
to simplify the fraction**
÷ by the same factor
top and bottom

$$\frac{6}{12} \frac{\div 2}{\div 2} \qquad \frac{3}{6} \frac{\div 3}{\div 3} \frac{1}{2}$$

Equivalent fractions
Take up the same space.

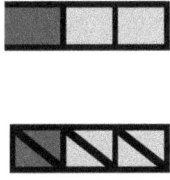

$\dfrac{1}{3}$

$\dfrac{2}{6}$

$\dfrac{1}{2} = \dfrac{2}{4}$ 3

$\dfrac{2}{6} \overset{\div 2}{\underset{\div 2}{}} \dfrac{1}{3}$

$\dfrac{1}{3} \div \dfrac{2}{6}$ 3
                    9

$\dfrac{1}{5} = \dfrac{2}{10}$ 3

To do equivalent
fractions ...

× top   numerator
× bottom   denominator

by the same number
factor

÷ by the same factor

Help funny Frank Fraction to complete the equivalent fractions.

$\dfrac{1}{2}$ $\dfrac{2}{4}$ $\dfrac{3}{6}$

$\dfrac{1}{3}$ $\dfrac{2}{6}$

$\dfrac{1}{4}$ $\dfrac{2}{8}$

$\dfrac{1}{5}$

$\dfrac{1}{6}$

$\dfrac{1}{8}$

Cancel down to their lowest form.

$\dfrac{\cancel{6}\ ^{\div 6}}{\cancel{12}\ _{\div 6}} = \dfrac{1}{2}$

$\dfrac{4}{8}$ $\dfrac{9}{12}$ $\dfrac{7}{21}$ $\dfrac{14}{21}$ $\dfrac{8}{12}$

Whole numbers

$\dfrac{15}{3}$ $\dfrac{8}{4}$ $\dfrac{12}{6}$

45

Find the fractional parts

$\dfrac{7}{8}$ of 56

$56 \div 8 = 7$
$7 \times 7 = 49$

$\dfrac{6}{8}$ of 48

$\dfrac{3}{4}$ of 60

$\dfrac{8}{9}$ of 72

$\dfrac{1}{2}$ of 14

÷ by bottom factor
(denominator)
× by top factor
(numerator)

Place in order from smallest:

$\dfrac{1}{2}$ $\dfrac{7}{12}$ $\dfrac{3}{4}$ $\dfrac{1}{6}$ $\dfrac{5}{12}$

Make the denominator the same value:

$\dfrac{1}{2}$ × $\dfrac{6}{6}$ $\dfrac{6}{12}$

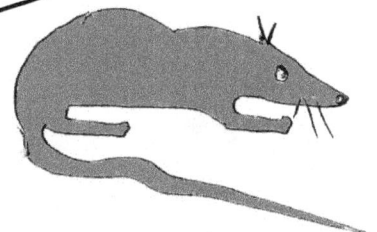

FRACTION SKILLS

0.5 is a decimal fraction or ÷ 5

$$\frac{5}{10} = \frac{1}{}$$

$\frac{5}{10}$ is a fraction

$$\frac{5}{10} = \frac{1}{} \text{ or } 0.5$$

Special ones

0.5 =

0.25 =

0.75 =

0.1 =

0.01 =

What is $\underset{\text{unit}}{} 2.4 \underset{\text{tenths}}{}$ as a fraction? $2^{4/10}$

What is **2.6** as a fraction?   Answer =

2.8 =

3.4 =

$$\frac{1}{2} = 0.5$$

$$\frac{1}{4} =$$

$$\frac{3}{4} =$$

$$\frac{1}{10} =$$

What is $5^{4/10}$ as a decimal?   **5.4**

$3\frac{6}{10} = 3.$ $9\frac{1}{10} =$

$2\frac{1}{10} =$ $6\frac{2}{5} =$

$4\frac{3}{5} =$

FOLLOW FRANK FRACTION'S EXAMPLE

<u>0.3</u> as a fraction is <u>3/10</u> or **30/100 = 30%** or **0.3**

0.6 =

0.9 =

0.2 =

0.4 =

4/5 as an equivalent fraction = 8/10 = 80/100 = 80%
4/5 as a decimal fraction = 0.8
÷ by 100, decimal point moves 2 steps to the left.

$$\frac{2}{5} = \frac{}{10} \quad \frac{}{100} \quad \% \qquad \frac{6}{20} =$$

$$\frac{7}{10} = \qquad\qquad\qquad \frac{3}{4} =$$

$2\dfrac{2}{5}$ $\dfrac{12}{5}$

is a mixed number

is an improper fraction

$2 \times 5 + 2 =$

$1\dfrac{4}{5} = \dfrac{9}{5}$

$2\dfrac{2}{3} =$

$6\dfrac{3}{4} =$

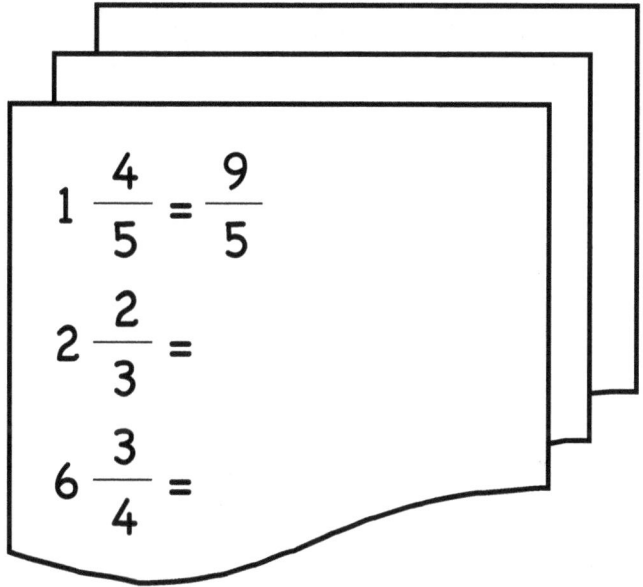

$\dfrac{7}{4} =$

$\dfrac{8}{3} =$

$\dfrac{6}{4} =$

$\dfrac{8}{3}$ is an improper fraction.

How many 3s in 8?

2 and 2 left $= \quad 2\dfrac{2}{3}$

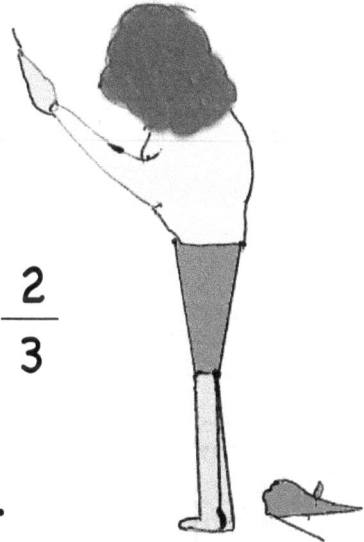

Make up some more.

Cancel down to the lowest form.

÷ top and bottom by the same factor.

$\dfrac{8}{40}$ $\dfrac{9}{27}$ $\dfrac{4}{36}$ $\dfrac{12}{36}$ $\dfrac{70}{100}$ $\dfrac{95}{100}$ $\dfrac{55}{100}$

HELP DARING DONNA DECIMAL RIDE THE COURSE ...

Match the **Fractions** to the **Decimals**.

$1 \frac{1}{4}$

2.75

$3 \frac{6}{10}$

5.3

$7 \frac{8}{10}$

$5 \frac{3}{10}$

match

1.25

$2 \frac{3}{4}$

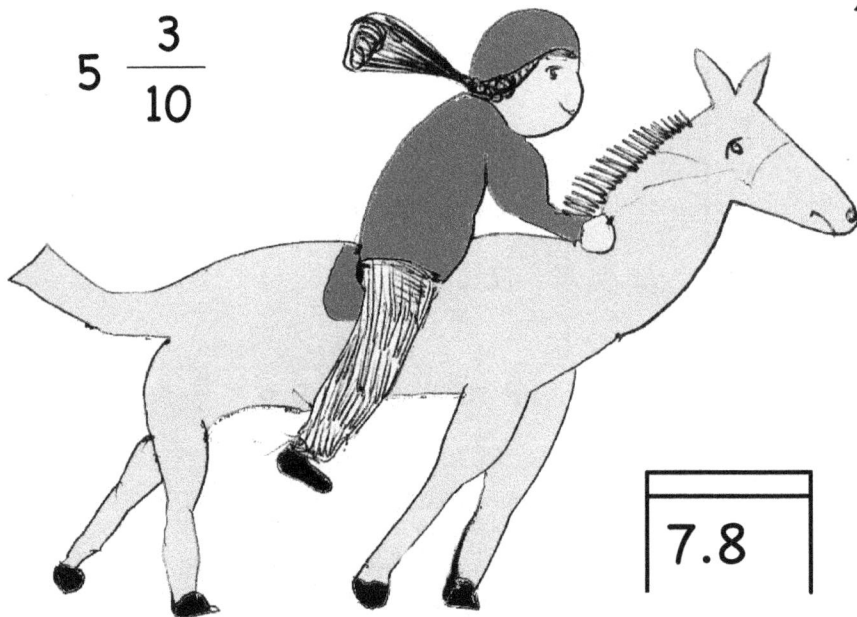

$14 \frac{1}{2}$

4.5

7.8

1	2	3	.	4	5

decimal point

Hundreds Tens Units Tenths Hundreds

3.6

$7 \frac{8}{10}$

14.5

$4 \frac{1}{2}$

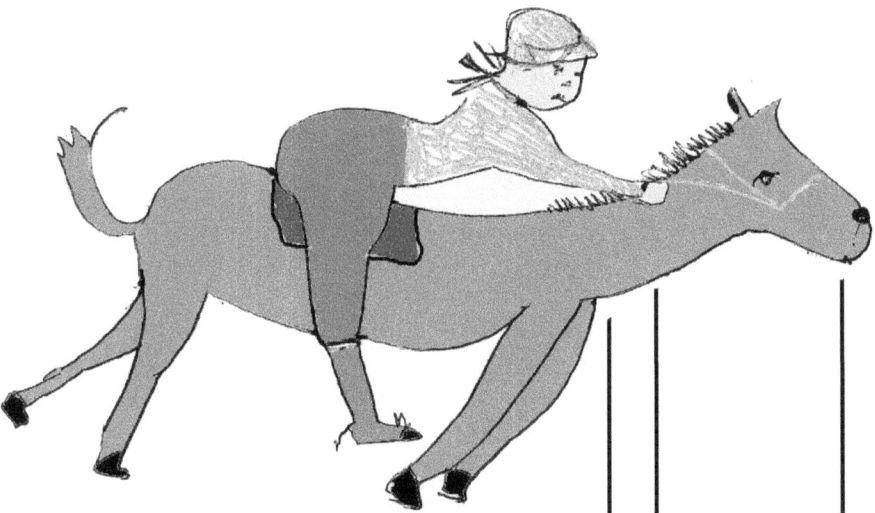

$\dfrac{7}{10}$ as a decimal = 0.7

$\dfrac{5}{100}$ as a decimal = 0.05

Tens	Units	.	Tenth	Hundredths
T	**U**	**.**	**T**	**H**
1	3	.	2	5

Match the decimals to the right fraction.

0.1

0.01

0.5

0.25

0.75

$\dfrac{25}{100}$ or $\dfrac{1}{4}$

$\dfrac{5}{10}$ or $\dfrac{1}{2}$ $\dfrac{1}{10}$ $\dfrac{1}{100}$

$\dfrac{75}{100}$ or $\dfrac{3}{4}$

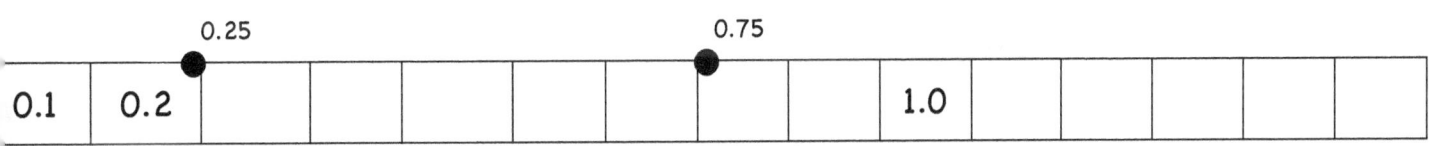

	0.25				0.75								
0.1	0.2							1.0					

10

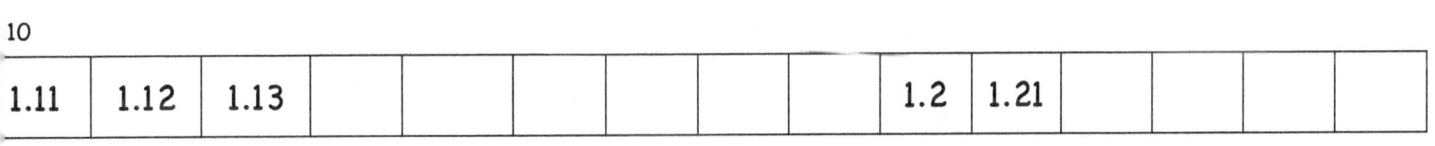

1.11	1.12	1.13						1.2	1.21				

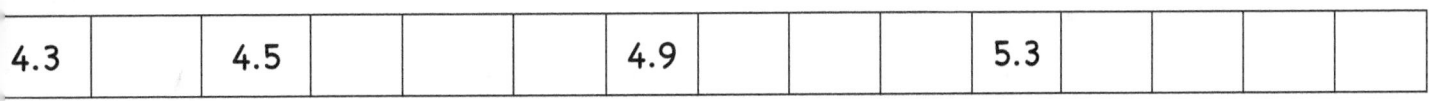

4.3		4.5			4.9			5.3				

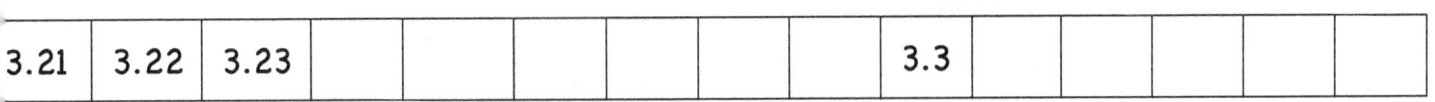

3.21	3.22	3.23						3.3					

53

1   4   .   1   2   3

to one decimal place   = 14.1

+ 1 here   over 5

1   4   .   ²1   6   3 ← ignore this time

to one decimal place   = 14.2

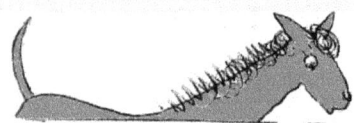

Donna has 10 riding lessons a term at £23.00
How much does her mum pay?

Joe has 100 driving lessons at £25.00 an hour. How much? (move decimal point 2 spaces.)

Move decimal point one space ⟶ $23.00 \times 10 =$

Her friend had 10 riding lessons at £18.00.
How much does her mum pay?

Another friend does not like horses. She has 12 flute lessons at £17.50 an hour.

How much does her mum pay?

To do long multiplication take out the point.

17.50 x 10
17.50 x 2
total:

Maths Practice

$$
\begin{array}{r}
1\ 7\quad 5\ 0 \\
\times \qquad 1\ 2 \\
\hline
3\ 5\quad 0\ 0 \\
1\ 7\ 5\quad 0\ 0 \\
\hline
2\ 1\ 0\ .\ 0\ 0 \\
\hline
\end{array}
$$

Add a zero

How many digits after the point?

Count the digits to the right of the decimal point and put the point back in.

Write these as percentages:

0.25 =

0.75 =

0.1 =

0.3 =

For example: 0.5

Percentages are always out of 100

$$\frac{5}{10} \begin{array}{c} \text{x10} \\ \\ \text{x10} \end{array} = \frac{50}{100} = 50\%$$

Money is in decimals.
Add Donna's bill at the pet shop.

HAY = £3.75
STRAW = £2.16
FOOD = £9.04

How much change from £20.00

You can just add a 0 so there are 2 digits after the .

5.08   4.17   5.8

3.20   3.45   3.21

Write them in order starting with the smallest.

Lisa has 18 lessons at £9.50 each.
How much does it cost?

Points together

4.92	£21.09	£15.73
+ 3.64	− £14.97	× 8

.   .   16.02

8) 330.00   6) 343.5   − 12.09

4.5 × 10 = 45   Move one point to the right to ×

3.6 × 10 =

0.2 × 10 =

19.2 ÷ 10 = 1.92   Move one point to the left to ÷

17.3 ÷ 10 =

4.6 ÷ 10 = 0.

Jo has 24 lessons at £10.50 each. How much does it cost?

Write as a fraction

0.5 =   0.75 =   0.25 =

0.1 =   0.01 =

$$\frac{1}{2} =$$

$$\frac{3}{4} =$$

$$\frac{7}{10} =$$

$$\frac{1}{4} =$$

$$\frac{3}{100} =$$

Build up these fractions

For example:

A percentage is out of 100

$$\frac{1}{5} \quad \frac{2}{10} \quad \frac{20}{100} \quad 20\% \quad 0.2$$

÷ 100 to find the decimal

$$\frac{2}{5}$$

$$\frac{3}{5}$$

$$\frac{4}{5}$$

$$\frac{5}{5}$$

$$\frac{1}{10}$$

$$\frac{2}{10}$$

7.6 x 100 =
21.7 x 100 =
0.3 x 100 =
(Move two points to the right to x)

53.2 ÷ 100 =
3064.5 ÷ 100 =
7.6 ÷ 100 =
(Move two points to the left to ÷)

$$0.1 \quad \frac{1}{10}$$

$$0.01 \quad \frac{1}{100}$$

30 x 10 =
(Add a 0 to x a whole number by 10.)
225 x 100 =
(Add two 0's to x a whole number by 10.)

450 ÷ 10 =
(Take off one 0 to divide a whole number by 10.)

7200 ÷ 100 =
(Take off two 0's to divide a whole number by 100.)

Can you complete the pattern?

$\frac{3}{4}$ ← numerator

4 ← denominator

$\frac{3}{4}$ of 40

$(40 \div 4) \times 3 =$

$10 \times 3 = 30$

0.75 is made up of $\frac{7}{10} = \frac{75}{100}$

0.625 is made up of $\frac{6}{10} = \frac{62}{100} = \frac{625}{1000}$

REVISE: **To find a percentage**

What is 30% of £50?

10% of 50 = 5

OR

30% = 5 × 3 = £15

$\frac{30}{100} \times \frac{50}{1} =$

$30 \div 100 \times 50$

What is 30% of £20?

10% of 20
30% of 20

Use **mental arithmetic** to work out these.

What is 20% of 30?

What is 40% of 60?

What is 80% of 70?

Perfect Patsy Percentage's

Work out the sale price of these items.

Shopping List		
Original Price	Saving	Sale Price
Bags and all things Beautiful Bag £30.00 SAVE 20%	$\frac{20}{100} \times \frac{30}{1} = £6$	£30 - £6 = £24
Shoes Galore Trainers £80.00 SAVE 25%		
Casual Clobber Trousers £50.00 SAVE 30%		
Fashion accessories Hairband £5.00 SAVE 10%		
Pet Place Rabbit food £4.50 SAVE 20%	Change £4.50 to 450p	
Fun Fashion Swimming costume £9.00 SAVE 10%		

Check your answers on a calculator.

e.g. 25 ÷ 100 x 80 = the cost of the shoes

Hint: change the pounds to pennies

Poppy's Place or Bonny's boutique

Which one shall I buy?

Use a calculator

$$\frac{\% \text{ off}}{100} \times \frac{\text{Amount of money}}{1} =$$

Blue trousers 75 ÷ 100 x 140 =   Green trousers 5 ÷ 100 x 60 =

Trousers

£140.00
75% off

£60.00
5% off

Which pair of trousers are the cheapest?
Which pair of shorts are the cheapest?
Which t-shirt is the cheapest?
Which pair of shoes are the cheapest?

£22.00
10% off

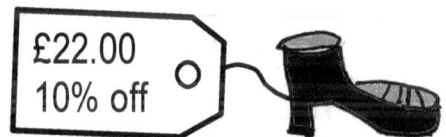

£55.00
5% off

£45.00
60% off

Shoes

£120.00
60% off

Shorts

£120.00
80% off

£25.00
5% off

T-shirts

Great stuff!

Which item is the best value for money?

Poppy's Place Designer Clothes

50% SALE £75 per dress

Bonny's Boutique

£120 £80

25% OFF

Bag's & Things

£30 per bag

20% off

SHOES GALORE

Up to 75% off

Patsy Percentage Shopaholic's sale

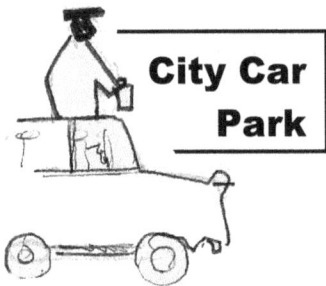

City Car Park

LOTTERY WINNER

Calculate the sale prices

TARIF

1 HOUR	90p
2 HOUR	£1.80
3 HOUR	£3.60
4 HOURS OR MORE	£4.50

ELEC DISCOUNT

HUGE SAVINGS PRICES SLASHED 60% OFF

Widescreen T.V, Video & D.V.D £1500.00

Garden World

40% reduction in store

Patio Set £350.00

PATSY'S PARKING PROBLEM

It is 10:20. Patsy leaves for home at 12:30.

How much money does she need for the meter?

10:20	→	11:00	→	12 :00	→		→	

Hairy Designer Hair stylist

35% off TODAY

Cut & blow £20.00

Casual Clobber £30

40% off

Fashion Accessories

£4.50

£3.50

10% off

PET PLACE

£40.00

20% SALE

Toys R U

15% off

Bear £20.00

58

Patsy's parents win

How do they spend their money? Look at the shopping list?

Find the shop on page 59 and work out the sale prices.

Shopping List

1) Wide screen T.V

2) Bear

3) Patio Set

4) Dress from Poppy's Place

5) Cut & Blow Dry

For example:

Wide Screen T.V

Shop: Elec. Discount

Cost: £1500

Discount: 60%

Saving: $\dfrac{60}{100} \times \dfrac{1500}{1} = 900$

$60 \div 100 \times 1500 = 900$

Sale Price: £1500 − £900 = **£600**

How much do they spend in total?

Do they have any change left from their lottery win?

To Find a Percentage

What percentage do these children score in tests?

Joel scored 20 marks out of 50 in a test. What is his mark as a percentage?

$$\frac{20}{\cancel{50}_1} \times \frac{\cancel{100}^2}{1} = 40\%$$ divide by 50

| Write the score as a fraction and times it by 100. |

$20 \div 50 \times 100 =$

These were Tara's marks in her summer exams. Find her marks as a percentage:

maths $\frac{30}{50}$ $\frac{30}{50} \times 100 = 30 \div 50 \times 100 = \quad \%$

English $\frac{80}{100}$

history $\frac{25}{40}$

geography $\frac{15}{20}$

science $\frac{30}{40}$

French $\frac{16}{20}$

Miss Melissa maths may be mean but not to her five cats that have all had kittens.

	Number of kittens
Millie	5
Mopsy	3
Molly	4
Maddy	7
Martha	6
Moa	5

Melissa maths cats have a total of:

$$5 + 7 + 3 + 4 + 6 + 5 = 30 \text{ kittens}$$

Melissa divides the total by the number of cats to find the **MEAN** or **AVERAGE**:

$$30 \div 6 = 5 \qquad \text{Mean} = 5$$

To find the **MODE** she puts them in order:

$$3 \quad 4 \quad 5 \quad 5 \quad 6 \quad 7$$

is the one that occurs most often

To find the **MEDIAN** she finds the middle number:

$$3 \quad 4 \quad 5 \quad 5 \quad 6 \quad 7$$

5 is the middle number

To find the **RANGE** take the smallest from the largest:

$$7 - 3 = 4$$

Miss Melissa Maths may not be so mean...

Total the items and divide by the number there are

Millie

5 Kittens

5 Kittens

Maddy

7 Kittens

Molly

4 Kittens

Martha

6 Kittens

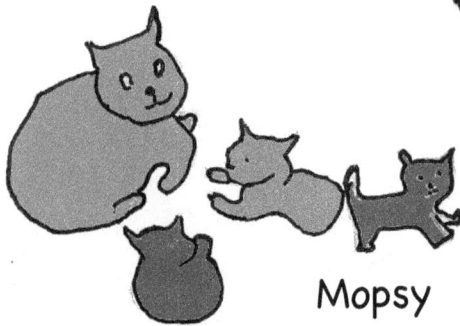

Mopsy

3 Kittens

Mean! What is that?

What is the mean number of hours
each adult cat sleeps each day?

Hint: Find total of
hours and ÷ 6.

Hours each adult cat
sleeps per day

Milly	Molly	Mopsy
12	15	9
Maddy	Martha	Moa
18	10	14

Each kitten eats 100g of food each day but their mums eat
200g each day.

How much food is consumed each day by each cat family?
(Go back to page 63.)

Use a calculator here.
Answer to one decimal place.

Maddy	900g	Moa		
Millie		Molly		Total
Mopsy		Martha		=		g		kg

What is the mean amount of food eaten?

100g packets of cat food cost 40p a packet. How much does
it cost each day to feed all the cats? Each week?

Melissa Maths Mean Problems...

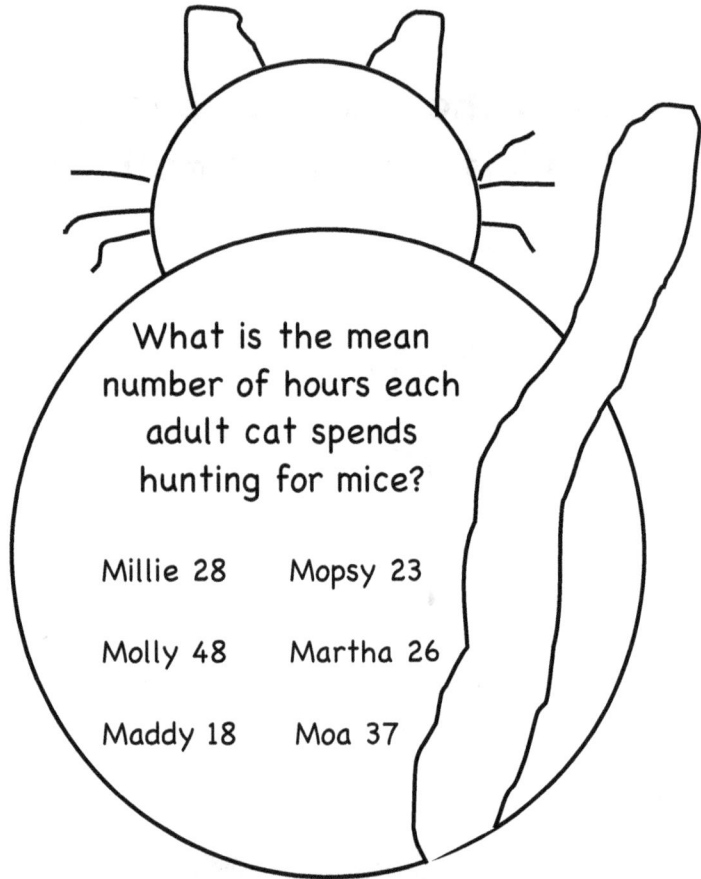

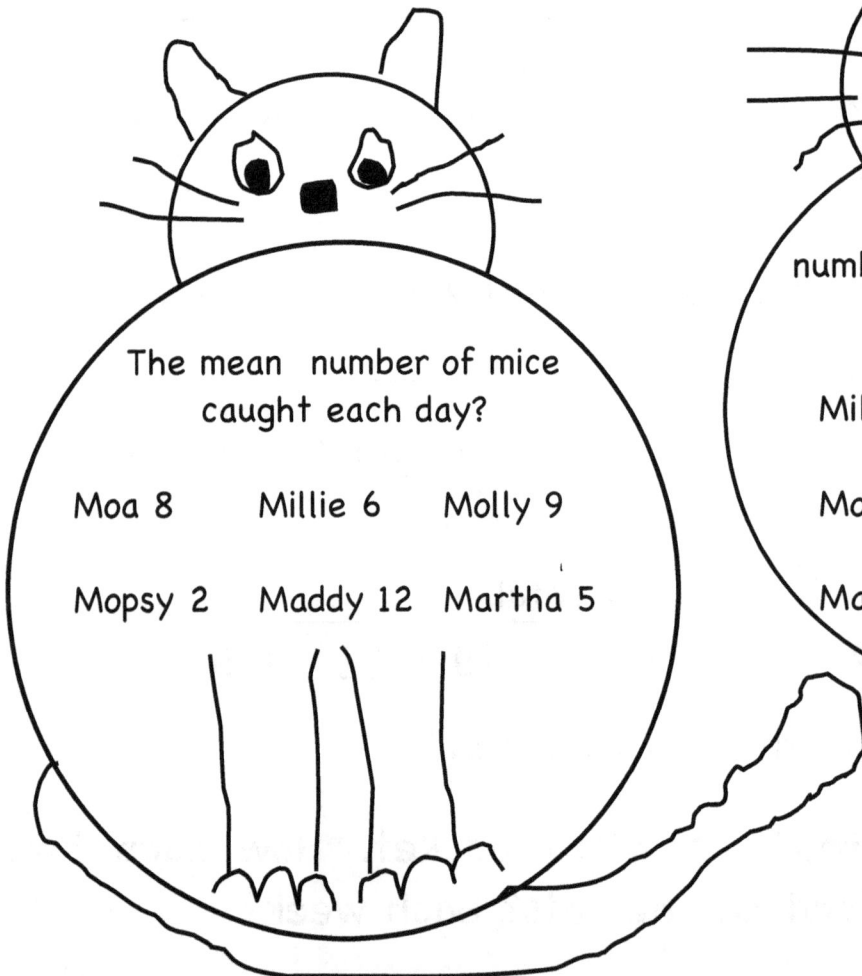

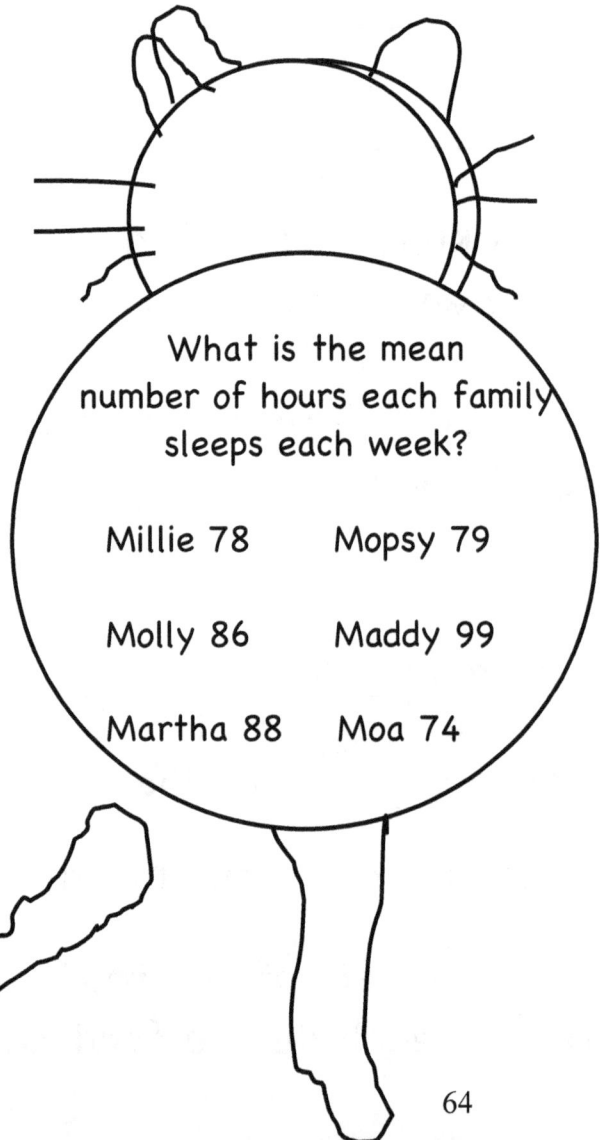

What is the mean size of a cat family including mum and kittens?

What is the mean number of hours each adult cat spends hunting for mice?

Millie 28   Mopsy 23

Molly 48   Martha 26

Maddy 18   Moa 37

The mean number of mice caught each day?

Moa 8   Millie 6   Molly 9

Mopsy 2   Maddy 12   Martha 5

What is the mean number of hours each family sleeps each week?

Millie 78   Mopsy 79

Molly 86   Maddy 99

Martha 88   Moa 74

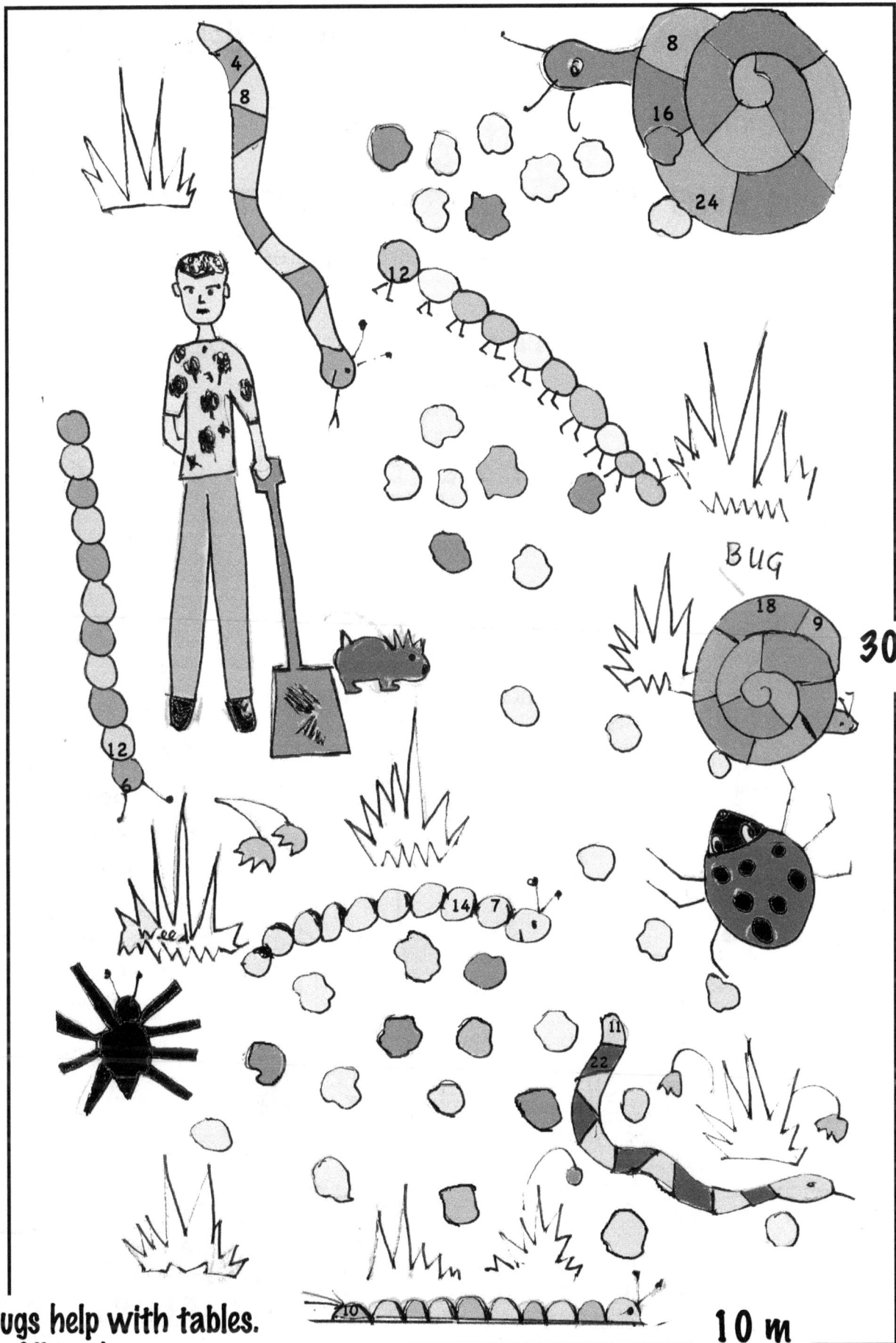

Angry Andrew argues over area and perimeter.
Find the area of Angry Andrews Garden

30 m

10 m

Bugs help with tables.
So fill in the sequences.

Angry Andrew finds the area, in square metres, of his garden.

Remember,

LENGTH X WIDTH = METRES ^{2}

30 m x 10 m = 300 metres squared

Angry Andrew adds a flowerbed 2.5m by 5m.

What is the area of the flower garden?

What area is left?

Remember: Answer in M^2

What is the perimeter of Andrews garden?

Perimeter of flowerbed =

Angry Andrews pal Paul Perimeter lends a hand.

PERIMETER is ALL THE WAY AROUND.

There are .... centimetres in 1 metre

There are .... centimetres in $\frac{1}{2}$ a metre

There are .... centimetres in $\frac{1}{4}$ of a metre

There are .... centimetres in $\frac{3}{4}$ of a metre

Help Angry Andrew calculate the area

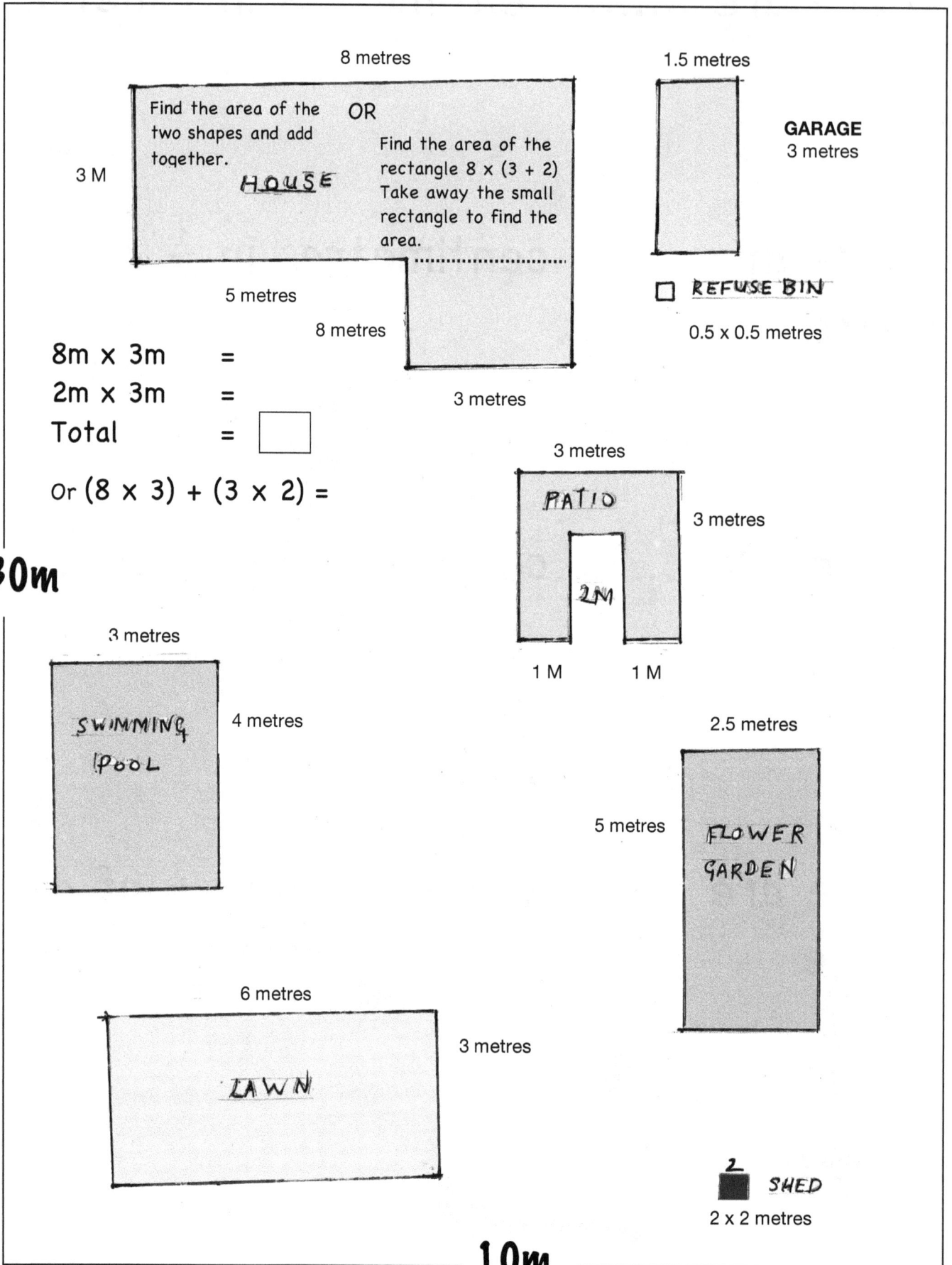

8 metres

Find the area of the two shapes and add together.   OR

HOUSE

Find the area of the rectangle 8 x (3 + 2) Take away the small rectangle to find the area.

3 M

5 metres

8 metres

1.5 metres

GARAGE
3 metres

□ REFUSE BIN

0.5 x 0.5 metres

8m x 3m   =

2m x 3m   =

Total   =

Or (8 x 3) + (3 x 2) =

3 metres

3 metres

PATIO

3 metres

2M

1 M   1 M

30m

3 metres

SWIMMING POOL

4 metres

2.5 metres

5 metres

FLOWER GARDEN

6 metres

3 metres

LAWN

2
SHED

2 x 2 metres

10m

Help Paul Perimeter add up the perimeter.

These guys say,

Remember

1000 grams = 1 kilogram

Katy Kilogram bakes a cake for 8 people.

She uses

100g sugar	0.1 kg
100g margarine	0.1 kg
100g flour	0.1 kg
2 eggs	

There are 24 children in her class. What ingredients will she need to make cakes for all of them?

How heavy is her 3.5 kg cake in grammes?

Thomas Tonne

1000 kilograms = 1 tonne

45 x 1000 = Add 3 zeros to a whole number.

Which is lighter?

2.3 kg x 1000
Move decimal point three places to the right = 2300 grams.

4065 grams ÷ 1000
Move decimal point three places to the left = 4.065 kg

Now try these ...

in g	in kg	in kg
2.4 kg =	4.2 t =	5023g =
5.72 kg =	1.1 t =	1203 g =
4.003 kg =	6.52 t =	72 g =

Megan millimetre

10 millimetres (mm) = 1 centimetre (cm)
1000 millimetres = 1 metre (m)

Milli means

$$\frac{1}{1000}$$

What is 75 mm in cm?   Answer 7.5 cm

What is 650 mm in cm?   Answer 65 cm

What is 650 mm in m?   Answer 0.65m

What is 45 mm in cm?
          in m?

Kind Kelly Kilometre and her friends

Can you answer their questions?

Remember

1000 metres (m)
= 1 kilometre (km)

Kilo means 1 thousand

What is 280m in km?   Answer 0.28 km

What is 230 m in km?   Answer ....

What is 250 cm in metres?

Answer 2.5m

What is 290 cm in metres?

Answer .....

Cilene centimetre says,

100 centimetres (cm)
= 1 metre (m)

Centi means

$$\frac{1}{100}$$

1000 millilitres (ml) = 1 litre (l)

1 kg   = ……………. grams

$\dfrac{1}{2}$ kg   = ……………. grams

$\dfrac{1}{4}$ kg   = ……………. grams

$\dfrac{3}{4}$ kg   = ……………. grams

Millimetre (mm)

Centimetre (cm)

Metre (m)

Kilometre (km)

1000 ml = 1 litre	
500 ml =	
250 ml =	
750 ml =	

1000g =	kg	250g =	kg
500g =	kg	750g =	kg

1000 Kg = 1 Tonne

1000 mm	=	m
10 mm	=	cm
100 cm	=	m
1000 m	=	km
1 km	=	m

71

Megan jumps 3.2m
How many cm? mm?

Cilene jumps 360cm
How many m? mm?

How much further
did Cilene jump?

Kelly jumps 1.4 m   140 cm
Cilene jumps 1.8 m   180 cm

How much further did
Cilene jump?

Answer in metres and centimetres.

Who ran the longest distance?
the shortest distance?

Sprint

Cilene ran	800 m	Kelly ran	200 m	Megan ran	40,000 cm
	cm		cm		m

Fill in the gaps.

1000g = 1kg

Gram	Kilograms
3400	
250	
4	
40	
1124	
3204	
506	
	0.07
	1.6
	9.2
	1.2
	2.4
	3.1
	7.24
	3.02
	5.001
	0.72
	0.010

Amit Angle
Adds up his triangles.

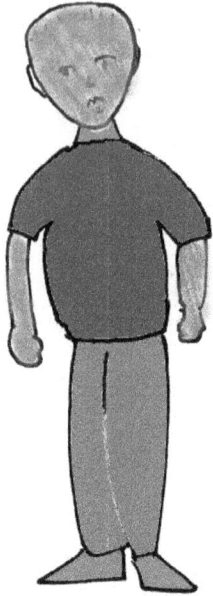

OBTUSE – a big fat angle, more than 90° less than 180°

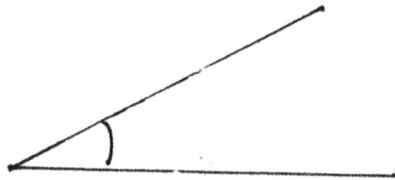

REFLEX – more than 180° less than 360°

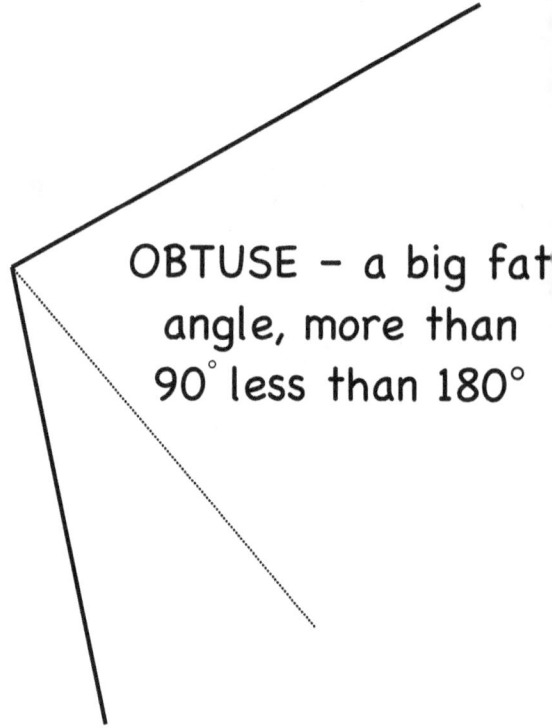

ACUTE or a cute little angle – less than 90°

Internal angles in a triangle add up to 180°

95°

42°

?

= 180°

180°

180°

$180 - (95 + 42) = 43°$

What is the missing angle?

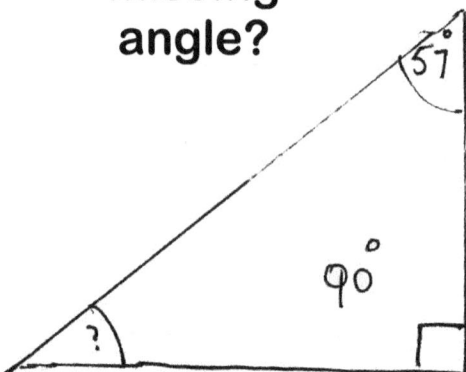

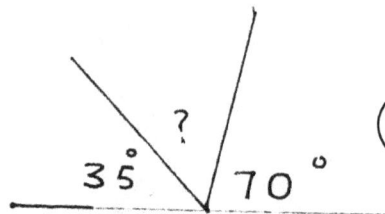

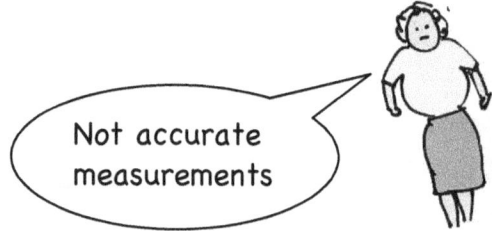

57°

90°

?

35° ? 70°

Not accurate measurements

Internal angles in a parallelogram add up to 360°

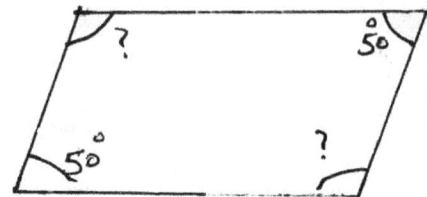

360°

? 50°

50° ?

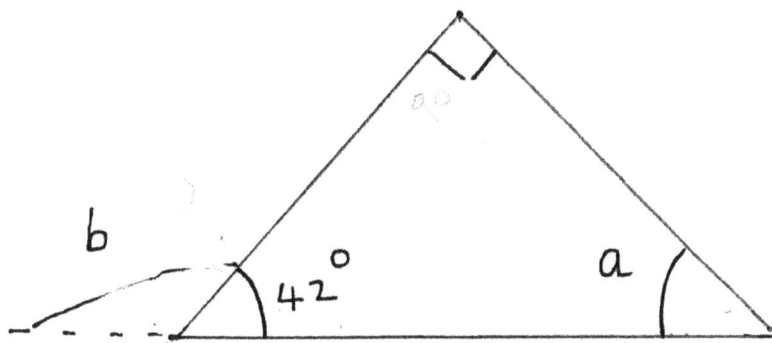

What is the right angle? = ☐ °

What do the angles in a triangle add up to?

Always ☐ °

What does angle a measure?

☐ + ☐ = ☐

☐ - ☐ = ☐

What does angle b measure?

A straight line measures ☐ °

☐ - ☐ = ☐

Find the missing angles.

a =   b =

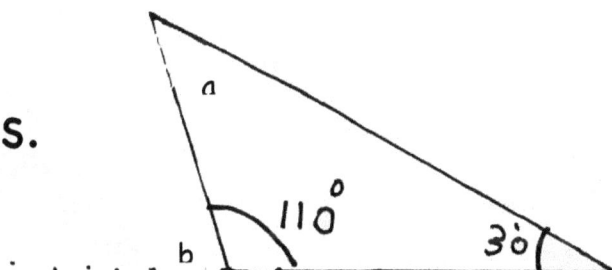

Match and Learn

Amit says a scalene triangle has NO SYMMETRY
NO EQUAL SIDES
NO EQUAL ANGLES

Which one?

Isosceles has
- symmetry = 1
- equal sides = 2
- equal angles = 2

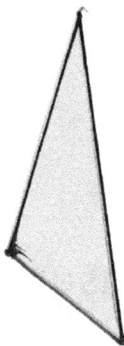

Equilateral triangle has
- symmetry = 3
- equal sides = 3
- angles of 60° = 3

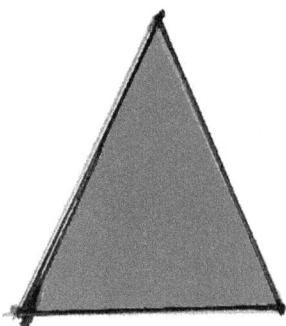

Reflex angles are more than 180° but less than 360°.

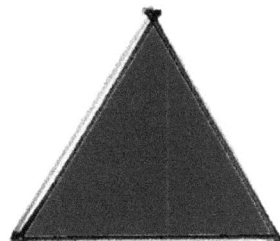

Right angle triangle

∟ 90°

Longest side hypotenuse

90°

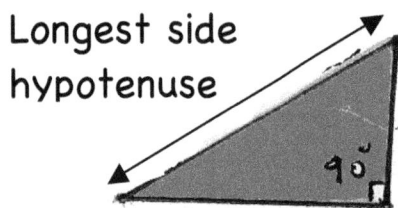

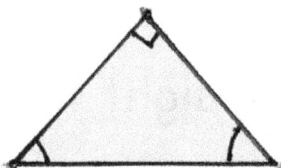

Acute angles are less than 90°

Obtuse angles are greater than 90° but less than 180°

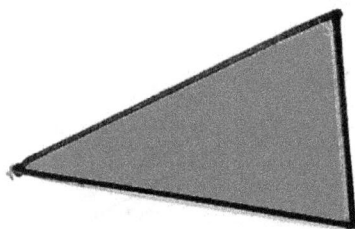

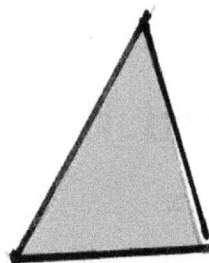

Get into shape with Charlie Cheerleader

Match the shapes to the correct information

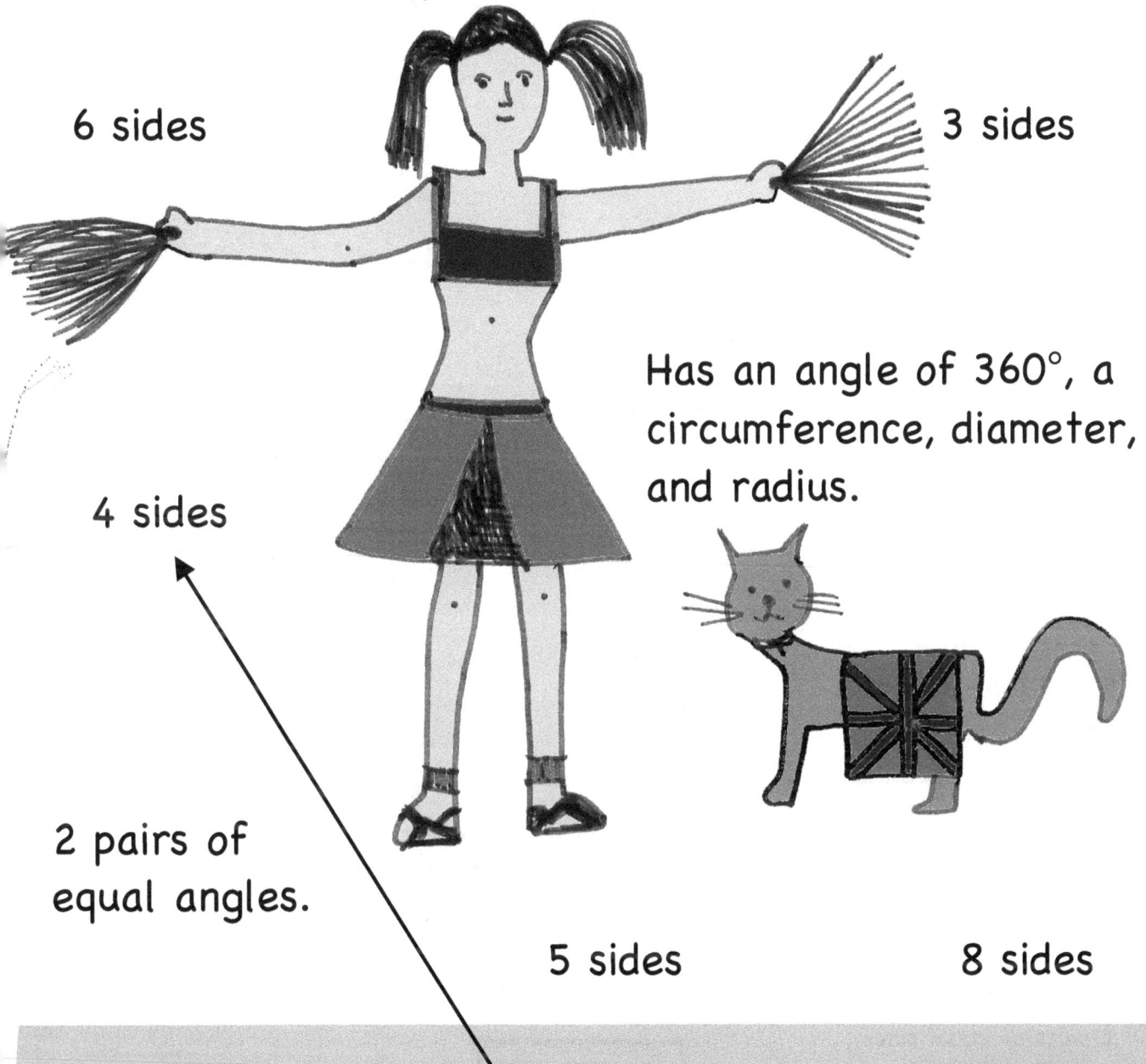

6 sides

3 sides

Has an angle of 360°, a circumference, diameter, and radius.

4 sides

2 pairs of equal angles.

5 sides

8 sides

| Pentagon | Square | Triangle | Circle |

| Hexagon | Octagon | Rhombus |

Types of Quadrilateral

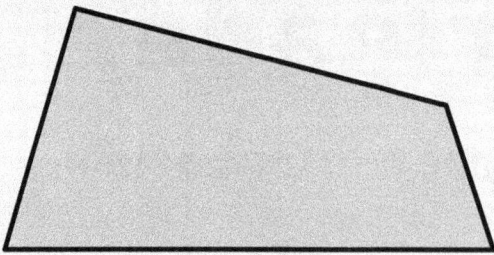

Quadrilateral is a **4 SIDED** shape. The interior **ANGLES ADD UP TO 360°.**

Rotational symmetry is how often the shape turns on itself.

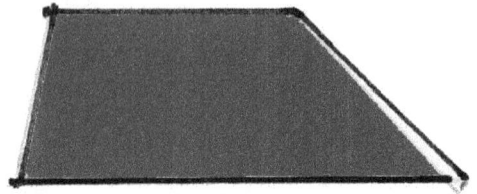

Square

- 4 equal sides
- 4 angles of 90°
- 4 lines of symmetry
- rotational symmetry order 4

Trapezium

- 1 pair of parallel sides
- no equal sides
- no equal angles
- no line of symmetry
- no rotational symmetry

Kite

- 2 pairs of adjacent equal sides
- 1 line of symmetry
- no rotational symmetry

Parallelogram

- 2 pairs of equal sides
- 2 pairs of parallel sides
- opposite angles are equal
- no lines of symmetry
- rotational symmetry order 2

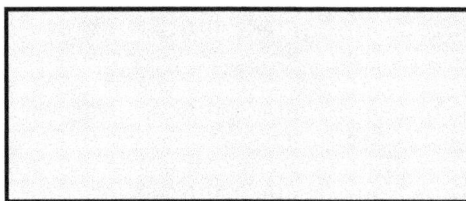

Rectangle

- 2 pairs of equal sides
- 4 angles of 90°
- 2 lines of symmetry
- rotational symmetry order 2

Rombus

- 4 equal sides
- 2 pairs of parallel sides
- 2 pairs of opposite equal angles
- 2 lines of symmetry
- rotational symmetry order 2

If the quadrilateral does not match one of these types it is an IRREGULAR QUADRILATERAL

78

Carry on the sequences

25   50   75   ....

250   500   750   1000   ....

60   120   180   ....

90   180   270   360   ....

$\frac{1}{2}$ 1 $1\frac{1}{2}$ 2 $2\frac{1}{2}$ 3   ....

$\frac{1}{4}$ $\frac{1}{2}$ $\frac{3}{4}$ 1 $1\frac{1}{4}$ $1\frac{1}{2}$ ....

Mental Addition with Decimals

$$4.6 + 8.16 = \cancel{12.22}$$

0 left off

Remember make both numbers into hundredths

$$4.60 + 8.16 = 12.76 \quad \checkmark$$

Add whole numbers

Add tenths   Add hundredths

3.8   +   8.17 =   5.2   +   3.42 =

7.1   +   5.62 =   15.4   +   8.45 =

13.06   +   3.5   =

Use Mental Arithmetic to solve multiplication problems

$$36 \times 8 = \quad \begin{array}{cc} 30 \times 8 = & 6 \times 8 = \\ 240 & 48 \end{array} \quad \begin{array}{l} \text{Total} \\ = 288 \end{array} \quad \begin{array}{r} 36 \\ \times\ 8 \\ \hline 2_28_48 \end{array}$$

Your turn:

34 × 5 =   63 × 3 =   41 × 4 =

46 × 6 =   76 × 2 =

Which two numbers have the difference of 19?

68 24

62 33

80 87

Look for units that can be subtracted to make 9

$$\begin{array}{r} {}^7\!\!\!\not{8}\;\;{}^1 7 \\ \underline{8} \end{array}$$

Ratio

Granny shared £30 between her 3 grandchildren. The eldest, Sophie had 3 times as much as baby Chloe. Joel had two times as much as Chloe. How much do they each get?

RATIO 3:2:1 = 6 shares (3 + 2 + 1 = 6)

£30 ÷ 6 = £5

Sophie = £5 × 3 = £15

Joel = £5 × 2 = £10

Chloe = £5 × 1 = £5

Now try these...

a) Grandad shared £150 between his 3 grandchildren. He gave Anna 3 times as much as toddler Tim and Sophie 2 times as much as Tim. How much do they each get?

b) Peter went into a shop and spent 4 times as much as Rachel while Kelly spent 2 times as much as Rachel. If they spent £49 between them, how much did they each spend?

Don't forget your negative numbers

-10 -9 -8 -7 -6 -5 -4 -3 -2 -1   0   1   2 3 4 5 6 7 8 9 10

Negative   Positive

Temperature goes up from   Temperature goes down from

-2   to   7°   = ☐°   5   to   -3°   = ☐°

-3   to   4°   = ☐°   7   to   -1°   = ☐°

-1   to   2°   = ☐°   3   to   -2°   = ☐°

3   to   12°   = ☐°   5   to   -1°   = ☐°

-3   to   5°   = ☐°   1   to   -1°   = ☐°

-8   to   11°   = ☐°   Make up some more
                                        examples

-14   to   8°   = ☐°

	0	×		0	=	3	0	0	0

Think of two factors that make 30.

Answer 6 and 5

$60 \times 50 = 3000$

	0	×		0	=	4	0	0	0

	0	×		0	=	4	2	0	0

Can you make up some more?

14:21

2:21 pm

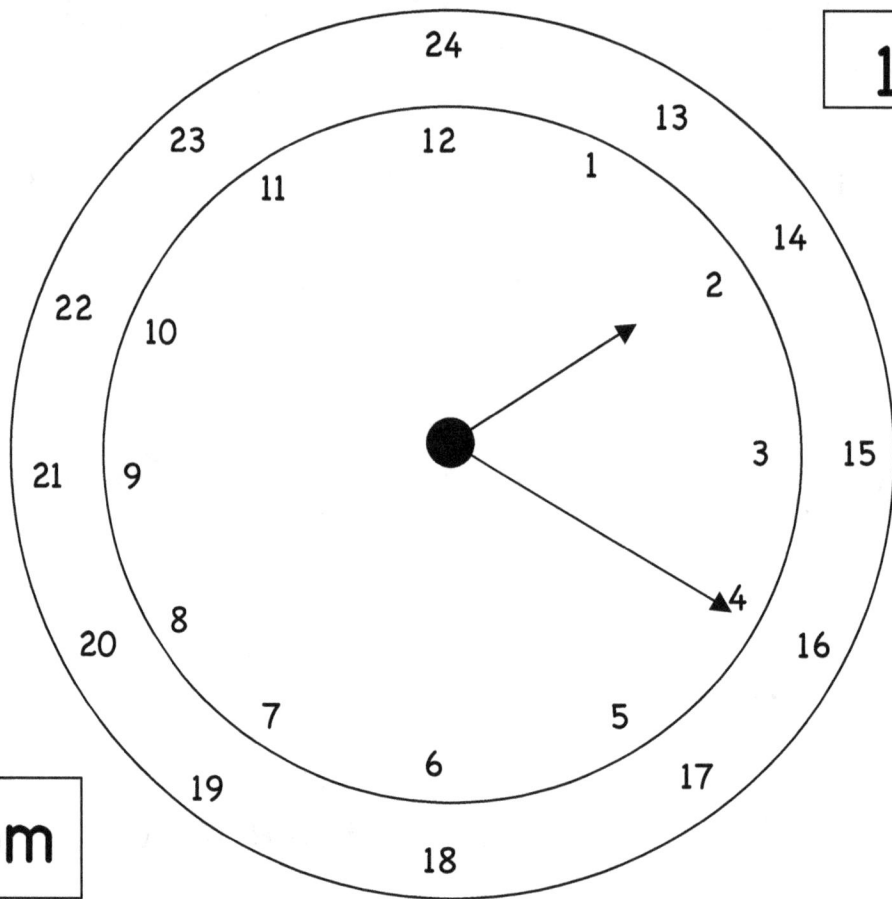

What is the Time?

12 HOUR CLOCK		24 HOUR CLOCK
12 hour midnight		00:00
4.30 am	=	04:30
12.00 noon midday		12:00
2.15 pm		14:15
11.59 pm		23:59

Write in 24-hour clock:

a) 7.15pm   b) 11.20pm

c) 9.30am   d) 3.15am

Write in analogue time:

a) 15 00   b) 18 30

c) 07 20   d) 10 15

Mathsland School Day Ends

$\dfrac{35}{50}$ in a test $=$ ___ %

Which is bigger

$\dfrac{3}{10}$ $\dfrac{4}{5}$

The difference between 55 and 39

The sum of 32 and 16 =

$8 \times 8 =$

$\dfrac{8}{100}$ as a decimal is 0.___

$\dfrac{4}{12}$ is equivalent to $\dfrac{1}{}$

Round 76 to the nearest 10
Round 252 to the nearest 100

$7\dfrac{1}{4}$ as a decimal $=$ ___

$\dfrac{80}{100} = \dfrac{}{10}$

0.___

10 more than 7296

1 m = ___ cm

4326
The value of the 2 =

3 less than 19.62

$\dfrac{1}{3}$
6

BACK HOME

5092
10 more =

$\dfrac{16}{5} = 3\dfrac{}{5}$

1 kg = ___ g

A multiple of 7

Circle the odd number

0.2 l = ___ ml

$0.48 \times 10 =$

The product of 9 and 8 =

10 less than 16.5

△ angles = ___ °

38

$6\dfrac{1}{2}$ hours ___ minutes

0.34×10

$4\dfrac{1}{3} = \dfrac{}{3}$

$38 \div 10 =$

47

4610
The value of 6 =

162

$\dfrac{4}{5} = \dfrac{8}{10} = \dfrac{80}{} = 80\%$

A factor of 54

The temperature rises from -5 to 4°. How many degrees?

Add 99 to 345

$\dfrac{2}{5} = \dfrac{4}{} = \dfrac{}{100} =$

0.7 + 0.3 + 21 =

$\dfrac{3}{4} = 0.$___

4.3 ___ $= 4\dfrac{3}{}$

219 ÷ 5 = 43r___

↑ Start YOUR JOURNEY HOME here!

Learn all these facts by filling in the gaps

Years in a Millennium ....

Years in a century ....

Years in a decade ....

Days in a year ....

Days in a leap year ....

The next leap year is in ....

There are .... months in a year

There are .... weeks in a year

There are .... days in a week

There are .... days in a fortnight

Complete the rhyme

30 days hath September, A................,
J................, N................. All the rest have 31 except
F................ which has 28 days clear but 29 in each
leap year.

What is today's date? ....

Write it the short way ....

What will the date be in 2 weeks time? ....

We are in the .... century.

1 minute = ... seconds

1 hour = ... minutes

1 day = ... hours

90° in a

180° in a _____

270° in a

360° in a circle

Fill in the points of the compass

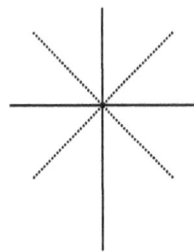

................grams = ... 1 kilogram

................grams = ... ½ kilogram

................grams = ... ¼ kilogram

................grams = ... ¾ kilogram

Test yourself

1) Write 1 million

2) 159
   675 +
   _____

   _____

3) 908
   66 -
   _____

   _____

4) 104
   8 ×
   _____

   _____

5) 4 ⟌ 260

6) 1 ½ kg = ………… g

7) Write in digits

   Eight thousand and twenty six

8) 67 p contains

   1 ………… p
   1 ………… p
   1 ………… p
   1 ………… p

9) 16:25 is the same as …………
   pm.

   quarter to five is 4.……… pm

10) or ………:……… in 24 hour
    clock.

11) A quadrilateral has …………
    sides.

12) 3 pieces of ribbon are cut from
    a metre. Each piece is 26 cm.
    How long is the piece left over?

13) $0.5 = \dfrac{1}{\text{……}}$

14) $0.25 = \dfrac{1}{\text{………}}$

15) $0.75 = \dfrac{\text{………}}{\text{………}}$

16) $0.1 = \dfrac{1}{\text{………}}$

17) $0.01 = \dfrac{1}{\text{………}}$

18) Write ¾ as a decimal.

19) Write ½ as a decimal.

20) Write ¼ as a decimal.

21) Write $\dfrac{2}{10}$ as a decimal.

22) 186 + ……… = 186
    186 − ……… = 186

23) A pentagon has ……… sides
    A hexagon has ……… sides
    An octagon has ……… sides

24) Write 1.7 as a fraction

25) What percentage of £5.00 is 50p?

26) What percentage of £1 is 20p?

27) The radius is half the diameter. If the diameter is 14 cm the radius is ………

28)

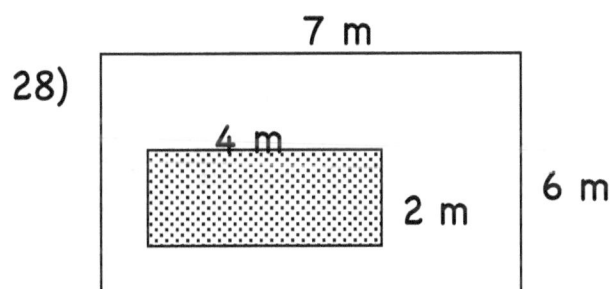

What area of grass is left after the flower bed is cut?

29) (6 × 7) + ……………. = 56

30) $4\dfrac{2}{5}$ as a decimal. (Make an equivalent fraction.)

31) ☐ seconds in 1 minute

    ☐ seconds in 2 minutes

    ☐ seconds in 3 minutes

32) 50 × 4 =

33) 11 22 33 ☐☐☐
    What is the rule?

34) 1 4 9 ☐☐☐
    What is the rule?

35) 3.6 × 10 =
    3.4 × 100 =

36) 60 − (7 × 8)

37) (36 ÷ 6) + 8 =

38) (9 × 7) − 10 =

39) 70 − (21 ÷ 3) =

40) 30 − (8 × 3) =

41)

	units	tenths	hundredth	thousandth
3.58				
14.25				
9.234				
18.05				
31.624				

42)

0.10   0.11   ...   .   0.2

3.01   ...   .   3.10

3.1   ...   .   3.2

43)

$4.5 \times 10 =$	$3.6 \times 10 =$	$72 \div 10 =$
$13.8 \div 10 =$	$31.24 \div 10 =$	$9.6 \div 10 =$
$5 \times 31 \times 10 =$	$71.5 \div 10 =$	$0.7 \div 10 =$

44)

$$\frac{}{12} = \frac{1}{4} \qquad \frac{6}{} = \frac{1}{2} \qquad \frac{6}{} = \frac{1}{3}$$

$$\frac{1}{3} = \frac{}{15} \qquad \frac{15}{20} = \frac{3}{} \qquad \frac{9}{} = \frac{1}{2}$$

45)

Triangular numbers

1, 3, 6, ...

46)

0.5 =   0.25 =   0.75 =   0.1 =   0.01 =

47)

```
    39                     82
 x  74                  x  46
 ---------              ---------
 .........              .........
     0                      0
 ---------              ---------
 _____                _____
```

```
         r                    r                    r
 3 | 436           4 | 387           9 | 808
```

```
         r                    r
 4 | 529           3 | 488
```

48)

4.36 x 100   19.1 ÷ 100

9.6 x 100   80 ÷ 100

2.31 x 100   71.5 ÷ 100

0.072 x 100   8.2 ÷ 100

49) Write in order. Start with the smallest.

5.35   2.65   5.7   2.05   5.03   19.5

20.05   19.29   21.6   18.46

50)

-16   -11   -6   -1   4   Rule?

....   0   -7   14   28   Rule?

Test yourself 2

1) 6 × 7 =   6 × 6 =   7 × 9 =   8 × 7 =

    6 × 8 =   7 × 6 =   6 × 9 =   7 × 8 =

    9 × 8 =   9 × 6 =   8 × 8 =   8 × 6 =

    7 × 7 =

2)

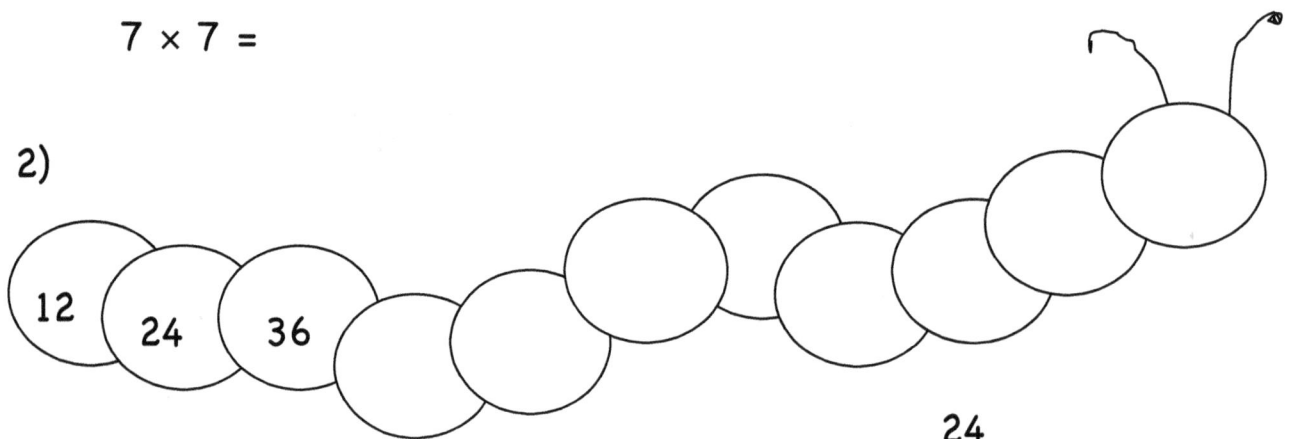

Count on 12 ...

3) 0.5 = $\dfrac{1}{\ldots\ldots}$

4) 0.25 = $\dfrac{1}{\ldots\ldots\ldots}$

5) 0.75 = $\dfrac{\ldots\ldots\ldots}{\ldots\ldots\ldots}$

6) 0.01 = $\dfrac{1}{\ldots\ldots\ldots}$

7) 0.2 = $\dfrac{\ldots\ldots\ldots}{\ldots\ldots}$

8) 0.7 = $\dfrac{\ldots\ldots\ldots}{\ldots\ldots}$

9) 0.46 = $\dfrac{\ldots\ldots\ldots}{100}$

10)

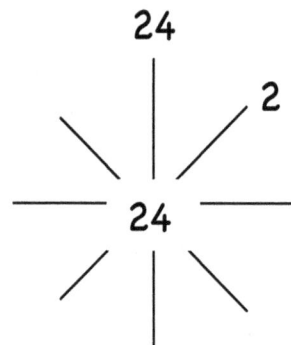

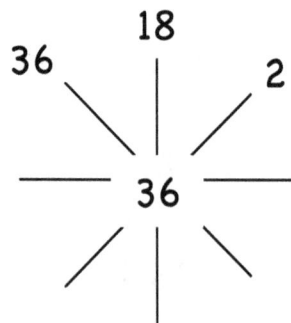

11) Write the prime numbers up to 30

    2   3   5 ....

92

12) Write ½ million

13) Write ¼ million

14) Write ¾ million

15) A five sided shape is called?
A six sided shape?
A seven sided shape?
An eight sided shape?

16) The area of a shape with length 7cm and width 6 cm is .... cm^2

17)

The name of this shape is?

18)

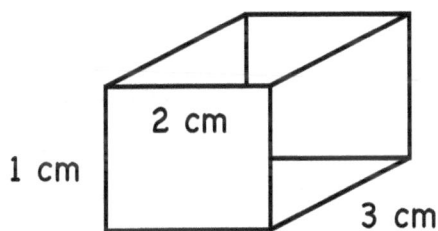

length x width x height

Find the volume

19) Build up these fractions

$\dfrac{1}{5}$ $\qquad$ $\dfrac{2}{10}$ $\qquad$ $\dfrac{20}{100}$ $\qquad$ 20% = 0.

20) $\dfrac{2}{5}$

21) $\dfrac{3}{5}$

22) $\dfrac{4}{5}$

23) Write the squared numbers

24) Write triangular numbers

25) $\dfrac{2}{10} = \dfrac{20}{100} = 20\% = 0.2$

$\dfrac{3}{10} = \dfrac{30}{100} = \qquad =$

$\dfrac{6}{10} = \dfrac{}{100} = \qquad =$

$\dfrac{4}{10} = \dfrac{40}{} = \qquad =$

26) Write down the multiples of 11

27) Write down the multiples of 12

28) $\dfrac{1}{2} =$

29) $\dfrac{1}{4} =$

30) $\dfrac{1}{3} =$

31) $\dfrac{1}{100} =$

32) $\dfrac{3}{4} =$

33) $\dfrac{1}{10} =$

34) $\dfrac{7}{10} =$

35) 1 kilogram = ... grams

36) ½ kg = $\qquad$ g

37) ¼ kg = $\qquad$ g

38) ¾ kg = $\qquad$ g

39) 500 ml = $\qquad$ l

40) 750 ml = $\qquad$ l

41) 250 ml =   l

42) 100 cm =   m

43) 1000 m =   km

44) Write a million

45) [   ] days in a week

[   ] months in a year

[   ] hours in a day

[   ] hours in 2 days

46) [   ] days in a year

[   ] days in a leap year

47)

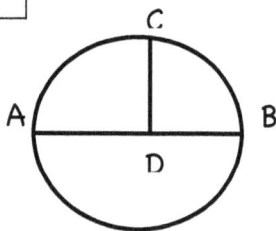

AB =   of a circle
CD =   of a circle

48) angles of a triangle add up to ...

49) straight line =   °

50) 176 x 0 =
149 x 1 =
136 ÷ 1 =
124 ÷ 0 =

51) [   ] + 7 = 17

[   ] + 9 = 23

[   ] − 8 = 12

[   ] − 4 = 16

52) 16:00 hours =

53) How many 10p coins in £1.90?

54) a − 7 = 10
a =

55) A train left at 10.25 am. The journey took 1 hour 20mins. What time did it arrive?

56) 9 squared =

57) 4364   3 stands for =

58) I took 6 hours to travel 300 miles. My average (mean) speed was ......

59) 40% of 50

60) If I save 50p a week. How many weeks will it take to save £20.00

61) Increase £104.75 by 66p

62) 36 + [   ] = 100

63)   > or <
   0.3   0.03

Answers

Page 10
a) 1,521,203
b) 965,428
c) 70,046
d) 5,138
e) 6,287,436
f) 605,245
g) 1,035,401
h) 376214 and 430,264
i) a) 149, 789, 846, 1,182,
6,032, 8,645
b) 1,236, 1,400, 7,060,
10,450, 100,000, 100,452

Page 11
8 + 9 = 17
7 + 8 = 15
5 + 13 = 18
12 + 4 = 16
6 + 8 = 14
3 + 14 = 17
5 + 15 = 20
13 + 2 = 15
14 + 5 = 19
10 + 6 = 16
11 + 7 = 18
17 + 2 = 19
9 + 10 = 19
6 + 7 = 13
5 + 15 = 20
11 + 9 = 20
12 + 6 = 18
7 + 5 = 12
17 + 3 = 20
14 + 6 = 20
4 + 8 = 12
12 + 5 = 17
7 + 7 = 14
10 + 6 = 16
9 + 5 = 14
10 + 8 = 18
13 + 6 = 19
1 + 18 = 19
18 + 1 = 19
9 + 9 = 18
14 + 1 = 15
7 + 6 = 13
6 + 5 = 11
14 + 6 = 20
9 + 6 = 15
13 + 5 = 18
10 + 2 = 12
9 + 5 = 14

Page 12
26 + 13 = 39
15 + 25 = 40
36 + 18 = 54
27 + 14 = 41

19 + 18 = 37
58 + 32 = 90
46 + 54 = 100
67 + 32 = 99
25 + 25 = 50
30 + 25 = 55
25 + 50 = 75
38 + 40 = 78
61 + 20 = 81
74 + 19 = 93
27 + 15 = 42
39 + 26 = 65
28 + 31 = 59
19 + 19 = 38
75 + 25 = 100
29 + 49 = 78
47 + 28 = 75
61 + 19 = 80
84 + 16 = 100
78 + 19 = 97
35 + 27 = 62
29 + 17 = 46
37 + 57 = 94
26 + 26 = 52
24 + 14 = 38
38 + 47 = 85
44 + 21 = 65
37 + 37 = 74

Page 15
6 × 6 = 36
6 × 7 = 42
6 × 8 = 48
6 × 9 = 54
7 × 6 = 42
7 × 7 = 49
7 × 8 = 56
7 × 9 = 63
8 × 6 = 48
8 × 7 = 56
8 × 8 = 64
8 × 9 = 72
9 × 6 = 54
9 × 7 = 63
9 × 8 = 72
9 × 9 = 81

2, 4, 6, 8, 10, 12, 14, 16, 18, 20
3, 6, 9, 12, 15, 18, 21, 24, 27, 30
4, 8, 12, 16, 20, 24, 28, 32, 36, 40
5, 10, 15, 20, 25, 30, 35, 40, 45, 50
6, 12, 18, 24, 30, 36, 42, 48, 54, 60
7, 14, 21, 28, 35, 42, 49, 56, 63, 70
8, 16, 24, 32, 40, 48, 56, 64, 72, 80
9, 18, 27, 36, 45, 54, 63, 72, 81, 90
10, 20, 30, 40, 50, 60, 70, 80, 90, 100

Page 17
1/4 = 0.25
1/2 = 0.5
3/4 = 0.75
1/10 = 0.1
1/100 = 0.01

3/10 30/100 30% 0.3
5/10 50/100 50% 0.5
6/10 60/100 60% 0.6
8/10 80/100 80% 0.8
9/10 90/100 90% 0.9

Page 18
60%

Page 20
3/4 of a circle = 270 degrees

- An equilateral triangle has 3 equal angles and 3 equal sides.
- An isosceles triangle has 2 equal angles and 2 equal sides.
- A scalene triangle has no equal sides or angles.

Page 21
1760 g   1.76kg
3 collars = £10.48
3 bowls = £10.40

Page 22
568 + 369 = 937
6692 + 4861 = 11,553
542 − 79 = 463
6478 − 4245 = 2,233
300 − 164 = 136
400 − 237 = 163
5368 + 1979 = 7,347
5000 − 1896 = 3,104
4000 − 3217 = 783
7359 + 1864 = 9,223
53 + 42 = 95
95 − 42 = 53
5 × 6 = 30
6 × 5 = 30
42 + 53 = 95
95 − 53 = 42
30 ÷ 5 = 6
30 ÷ 6 = 5

Page 23
The difference between 4037 and 2864 = 1,173
Subtract 3654 from 2391 = 1,263
5238 minus 3400 = 1,838
$3.2 - 2.6 = 0.6$
$4862 - 3000 = 1,862$
$4.83 - 2.40 = 2.43$
$40000 - 3954 = 36,046$

Page 24
$716 - 235 = 481$
$823 - 235 = 588$
$645 - 235 = 410$
$547 - 362 = 185$
$362 - 243 = 119$

Page 25
$85 - 35 = 50$
$70 - 25 = 45$
$531 - 259 = 272$
$934 - 869 = 65$
$5726 - 4487 = 1,239$
$67 - 23 = 44$
$61 - 18 = 43$
$511 - 294 = 217$
$548 - 374 = 174$
$5810 - 996 = 4,814$
$24 - 18 = 6$
$631 - 358 = 273$
$644 - 357 = 287$
$542 - 357 = 185$
$5327 - 4245 = 1,082$
$200 - 153 = 47$
$1000 - 834 = 166$
$600 - 423 = 177$
$4020 - 3153 = 867$
$7000 - 568 = 6,432$
$6003 - 428 = 5,575$

Page 26
6, 12, 18, 24, 30, 36, 42, 48, 54, 60, 66, 72.

15 1, 15, 3, 5
54 1, 54, 2, 27, 3, 18, 6, 9,
72 1, 72, 2, 36, 3, 24, 4, 18 6, 12, 9, 8
21 1, 21, 7, 3
36 1, 36, 2, 18, 3, 12, 4, 9, 6

Page 27
Prime numbers
2, 3, 5, 7, 11, 13, 17, 19, 23, 29, 31, 37, 41, 43, 47, 53, 59, 61, 67, 71, 73, 79, 83, 89, 97.

Page 28
23, 28, 33, 38, 43
Rule = +5
1, 3, 6, 10, 15
Rule = triangular numbers +1, +2, +3 etc. Add one more each time.
95, 89, 83, 77
Rule = -6
8, 16, 24, 32, 40
Rule = +8
1, 4, 9, 16, 25
Rule = square numbers 1x1, 2x2 etc.
70, 63, 56, 49, 35
Rule = -7.
7, 8, 10, 13, 17
Rule = +1 more each time.
8, 18, 28, 38, 48
Rule = +10.

Page 29
$10 \times 4 = 40$
$1 \times 2 = 2$
$8 \times 2 = 16$
$6 \times 4 = 24$
$3 \times 3 = 9$
$3 \times 2 = 6$
$7 \times 5 = 35$
$4 \times 4 = 16$
$3 \times 6 = 18$
$4 \times 6 = 24$
$5 \times 8 = 40$
$1 \times 5 = 5$
$9 \times 4 = 36$
$7 \times 3 = 21$
$9 \times 2 = 18$
$6 \times 5 = 30$
$3 \times 5 = 15$
$6 \times 2 = 12$
$10 \times 5 = 50$
$2 \times 7 = 14$
$3 \times 7 = 21$
$2 \times 2 = 4$
$6 \times 3 = 18$
$7 \times 4 = 28$
$7 \times 2 = 14$
$8 \times 3 = 24$
$10 \times 2 = 20$
$9 \times 3 = 27$
$9 \times 5 = 45$
$1 \times 4 = 4$
$7 \times 7 = 49$
$5 \times 7 = 35$
$3 \times 8 = 24$
$8 \times 4 = 32$
$2 \times 5 = 10$
$2 \times 2 = 4$
$10 \times 3 = 30$
$5 \times 4 = 20$
$8 \times 5 = 40$

$3 \times 4 = 12$
$1 \times 3 = 3$
$4 \times 7 = 28$
$6 \times 7 = 42$
$1 \times 9 = 9$
$4 \times 2 = 8$
$5 \times 3 = 15$
$2 \times 3 = 6$
$1 \times 6 = 6$
$1 \times 8 = 8$
$3 \times 9 = 27$
$2 \times 4 = 8$
$2 \times 9 = 18$
$1 \times 7 = 7$
$2 \times 6 = 12$
$4 \times 8 = 32$
$4 \times 9 = 36$

Page 30
$5 \times 11 = 55$
$1 \times 11 = 11$
seven times six = 42
multiply 9 by 6 = 54
$12 \times 11 = 132$
$2 \times 11 = 22$
$4 \times 11 = 44$
find the product of 8 and 6 = 48
$8 \times 11 = 88$
$6 \times 11 = 66$
$9 \times 11 = 99$
find the product of 12 and 8 = 96
$3 \times 11 = 33$
$10 \times 11 = 110$

$12 \times 4 = 48$
$12 \times 12 = 144$
$12 \times 10 = 120$
$12 \times 6 = 72$
$2 \times 12 = 24$
$12 \times 5 = 60$
$12 \times 11 = 132$
$12 \times 3 = 36$
$12 \times 8 = 96$
$12 \times 9 = 108$
$12 \times 7 = 84$
$1 \times 12 = 12$

12, 24, 36, 48, 60, 72, 84, 96, 108, 120, 132, 144

Page 31
1) 2, 3, 5, 7, 11, 13, 17, 19, 23, 29
2) 1, 4, 9, 16, 25, 36, 49, 64, 81, 100
3) 1, 8, 27, 64, 125, 216
4) 1, 3, 6, 10, 15, 21, 28, 36, 45, 55

5)

6 - 6, 12, 18, 24, 30, 36, 42, 48, 54, 60

8 - 8, 16, 24, 32, 40, 48, 56, 64, 72, 80

7 - 7, 14, 21, 28, 35, 42, 49, 56, 63, 70

9 - 9, 18, 27, 36, 45, 54, 63, 72, 81, 90

12 - 12, 24, 36, 48, 60, 72, 84, 96, 108, 120, 132, 144

6)

12 1, 12, 4, 2, 3, 6
15 1, 15, 3, 5
24 1, 24, 2, 12, 3, 8, 6, 4
36 1, 36, 2, 6, 9, 18, 3, 4, 12
18 1, 18, 2, 6, 3, 9

Page 32
$6 \times 6 = 36$
$6 \times 7 = 42$
$6 \times 8 = 48$
$6 \times 9 = 54$
$7 \times 6 = 42$
$7 \times 7 = 49$
$7 \times 8 = 56$
$7 \times 9 = 63$
$8 \times 6 = 48$
$8 \times 7 = 56$
$8 \times 8 = 64$
$8 \times 9 = 72$
$9 \times 6 = 54$
$9 \times 7 = 63$
$9 \times 8 = 72$
$9 \times 9 = 81$

$1 \times 1 = 1$
$2 \times 2 = 4$
$3 \times 3 = 9$
$4 \times 4 = 16$
$5 \times 5 = 25$
$6 \times 6 = 36$
$7 \times 7 = 49$
$8 \times 8 = 64$
$9 \times 9 = 81$
$10 \times 10 = 100$

6, 12, 18, 24, 30, 36, 42, 48, 54, 60

7, 14, 21, 28, 35, 42, 49, 56, 63, 70

15, 30, 45, 60, 75, 90, 105, 120, 135, 150

50, 100, 150, 200, 250, 300, 350, 400. 450, 500

250, 500, 750, 1000, 1,250, 1,500, 1,750, 2000, 2,250, 2,500

60, 120, 180, 240, 300, 360, 420, 480, 540, 600

90, 180, 270, 360, 450, 540, 630, 720, 810, 900

½, 1, 1 ½, 2, 2 ½, 3, 3 ½, 4, 4 ½, 5

¼, ½, ¾, 1, 1 ¼, 1 ½, 1 ¾, 2, 2 ¼, 2 ½, 2 ¾, 3

Page 33
$23 \times 6 = 138$
$29 \times 7 = 203$
$28 \times 4 = 112$
$426 \times 8 = 3,408$
$33 \times 8 = 264$
$325 \times 7 = 2,275$
$4372 \times 9 = 39,348$

$36 \times 4 = 144$
($30 \times 4 = 120$, $6 \times 4 = 24$, $120 + 24 = 144$)

$82 \times 6 = 492$
($80 \times 6 = 480$, $2 \times 6 = 12$, $480 + 12 = 492$)

$31 \times 5 = 155$
($30 \times 5 = 150$, $1 \times 5 = 5$, $150 + 5 = 155$)

Page 34/35
$63 \div 7 = 9$
7, 14, 21, 28, 35, 42, 49, 56, 63
$7 \times 9 = 63$
$9 \times 7 = 63$

Page 36
$14 \div 7 = 2$
$35 \div 7 = 5$
$28 \div 7 = 4$
$21 \div 7 = 3$
$42 \div 7 = 6$
$49 \div 7 = 7$
$56 \div 7 = 8$
$119 \div 7 = 17$
$133 \div 7 = 19$
$70 \div 7 = 10$
$77 \div 7 = 11$
$98 \div 7 = 14$
$105 \div 7 = 15$
$63 \div 7 = 9$
$7 \times 9 = 63$

Page 37
$69 \div 7 = 9$ r6

Page 38
$69 \div 8 = 8$ r5
$50 \div 7 = 7$ r1
$79 \div 4 = 19$ r3
$156 \div 6 = 26$
$84 \div 9 = 9$ r3
$59 \div 6 = 9$ r5
$146 \div 9 = 16$ r2
$64 \div 9 = 7$ r1
$490 \div 7 = 70$
$1036 \div 8 = 129$ r4
$39 \div 5 = 7$ r4
$104 \div 8 = 13$
$732 \div 5 = 146$ r2

Page 39
$16 \times 28 = 448$
$36 \times 24 = 864$
$125 \times 42 = 5,250$

Page 40
$640 \div 20 = 32$
$770 \div 35 = 22$
$240 \div 16 = 15$
$432 \div 18 = 24$
$336 \div 14 = 24$

Page 41
16, 32, 48, 64, 80, 96, 112, 128, 144, 160, 176, 192

$43 \times 19 = 817$
$45 \times 16 = 720$

Page 42
- £266 ÷ £19.00 = 14 weeks
- £204 ÷ £17 = 12 weeks
- £1408 ÷ 22 weeks = £64 per week
- £12 × 15 weeks = £180

17, 34, 51, 68, 85

22, 44, 66, 88

15, 30, 45, 60, 75

Page 43
14, 28, 42, 56, 70, 84, 98, 112, 126, 140

$48 \times 17 = 816$
$49 \times 14 = 686$
$14 \times 28 = 392$
$36 \times 15 = 540$

Page 45
Fraction of black rats = 2/5
Ate 10 = 10/16, 5/8
Ate 6 = 6/12, 1/2
Ate 12 = 12/18, 2/3
Ate 3 = 3/12 / ¼
Drank 5 = 5/8

Page 46
1/2, 2/4, 3/6, 4/8, 5/10, 6/12

1/3, 2/6, 3/9, 4/12, 5/15, 6/18

1/4, 2/8, 3/12, 4/16, 5/20, 6/24

1/5, 2/10, 3/15, 4/20, 5/25, 6/30

1/6, 2/12, 3/18, 4/24, 5/30, 6/36

1/8, 2/16, 3/24, 4/32, 5/40, 6/48

4/8 lowest form = 1/2
9/12 lowest form = 3/4
7/21 lowest form = 1/3
14/21 lowest form = 2/3
8/12 lowest form = 2/3
15/3 lowest form = 5
8/4 lowest form = 2
12/6 lowest form = 2

Page 47
6/8 of 48 = 36
3/4 of 60 = 45
8/9 of 72 = 64
1/2 of 14 = 7

1/6, 5/12, 1/2, 7/12, ¾
(3/4 = 9/12,
(1/2 = 6/12)

Page 48
0.5 = ½
0.25 = ¼
0.75 = ¾
0.1 = 1/10
0.01 = 1/100
½ = 0.5
¼ = 0.25
¾ = 0.75
1/10 = 0.1

2.6 as a fraction = 2 6/10
2.8 as a fraction = 2 8/10
3.4 as a fraction = 3 4/10

3 6/10 as a decimal = 3.6
2 1/10 as a decimal = 2.1
4 3/5 as a decimal = 4.6
9 1/10 as a decimal = 9.1
6 2/5 as a decimal = 6.4

Page 49
0.6 = 6/10 or 60/100 = 60%
0.9 = 9/10 or 90/100 = 90%
0.2 = 2/10 or 20/100 = 20%
0.4 = 4/10 or 40/100 = 40%

2/5 = 4/10, 40/100 = 40%   0.4
7/10 = 70/100 = 70%   0.7
6/20 = 3/10, 30/100 = 30%   0.3
3/4 = 75%   0.75

Page 50
2 2/3 = 8/3
6 3/4 = 27/4

7/4 = 1 3/4
8/3 = 2 2/3
6/4 = 1 2/4, 1 1/2

8/40 = 1/5
9/27 = 1/3
4/36 = 1/9
12/36 = 1/3
70/100 = 7/10
95/100 = 19/20
55/100 = 11/20

Page 51
5.3 – 5 3/10
2.75 – 2 ¾
1.25 – 1 ¼
4.5 – 4 ½
7.8 – 7 8/10
14.5 – 14 ½
3.6 – 3 6/10

Page 52
0.1= 1/10, 0.01= 1/100, 0.5 = ½
0.25= ¼, 0.75= 3/4

0.1, 0.2, 0.3, 0.4, 0.5, 0.6, 0.7
0.8, 0.9, 1.0, 1.1. 1.2, 1.3, 1.4, 1.5

1.11, 1.12, 1.13, 1.14, 1.15. 1.16, 1.17,
1.18, 1.19, 1.20, 1.21, 1.22, 1.23, 1.24,
1.25

4.3, 4.,4, 4.5, 4.6, 4.7, 4.8, 4.9, 5.0,
5.1, 5.2, 5.3, 5.4, 5.5, 5.6, 5.7

3.21, 3.22, 3.23, 3.24, 3.25, 3.26,
3.27, 3.28, 3.29, 3.30, 3.31, 3.32,
3.33, 3.34, 3.35

Page 53
• £230,00
• £180.00
• £210.00

Page 54
0.25 = 25/100 - 25%
0.75 = 75/100 - 75%
0.1 = 10/100 - 10%
0.3 = 30/100 - 30%

• £3.75 + £2.16 + £9.04 = £14.95
£5.05 change from £20.00

• 3.2, 3.21, 3.45, 4.17, 5.08, 5.80

• £9.50 x 18 = £171.00

4.92 + 3.64 = 8.56
£21.09 - £14.97 = 6.12
£15.73 x 8 = £125.84
330.00 ÷ 8 = 41.25
343.5 ÷ 6 = 57.25
16.02 – 12.09 = 3.93

3.6 x 10 = 36
0.2 x 10 = 2
17.3 ÷ 10 = 1.73
4.6 ÷ 10 = 0.46

• £10.50 x 24 =£252.00

Page 55
0.5 = 1/2
0.1 = 1/10
0.75 = 3/4
0.01 = 1/100
0.25 = 1/4

½ = 0.5
¾ = 0.75
7/10 = 0.7
¼ = 0.25
3/100 = 0.03

7.6 x 100 = 760, 21.7 x 100 = 2170
0.3 x 100 = 30, 53.2 ÷ 100 = 0.532
3064.5 ÷ 100 = 30.645, 7.6 ÷ 100 =
0.076

2/5, 4/10, 40/100, 40%, 0.4
3/5, 6/10, 60/100, 60%, 0.6
4/5, 8/10, 80/100, 80%, 0.8
5/5, 10/10, 100/100, 100%, 1.0
1/10, 10/100, 10%, 0.1
2/10, 20/100, 20%, 0.2

Page 56
30% of £20 = £6

20% of 30 = 6
40% of 60 = 24
80% of 70 = 56

Page 57
- Trainers £80   Save 25%

Saving = £20.00

Sale price = £60.00

- Trousers £50   Save 30%

Saving = £15.00

Sale price = £35.00

- Hairband £5.00   Save 10%

Saving = 50p

Sale price = £4.50

- Rabbit food £4.50   Save 20%

Saving = 90p

Sale price = £3.60 or 360 pennies.

- Swimming costume £9.00
   Save 10%

Saving = 90p

Sale price = £8.10

Page 58
Trousers
£140 with 75% off

Saving = £105.00

Sale Price = £35

(Best value for money.)

£60 with 5% off

Saving = £3.00

Sale Price = £57.

Shorts
£120 with 60% off

Saving = £72.00

Sale Price = £48.00

(Best value for money)

£55 with 5% off

Saving = £2.75

Sale Price: £52.25

T-Shirts
£120 with 80% off

Saving = £96.00.

Sale Price = £24.00

£25 with 5% off

Saving = £1.25

Sale Price = £23.75

(best value for money)

Shoes
£45.00 with 60% off

Saving = £27.00

Sale Price = £18.00

(Best value for money.)

£22.00 with 10%

Saving = £2.20

Sale Price = 19.80

Page 59

2 hours, 10 minutes.

Patsy needs £3.60 for the meter.

Page 60

Bear
Toys R U

Cost: £20.00

Discount: 15%

Saving = £3.00

Sale price = £17.00

Patio Set
Garden World

Cost: £350.00

Discount: 40%

Saving = £140

Sale price = £210.00

Dress
Poppy's Place

Cost: £75

Discount: 50%

Saving = £37.50

Sale price = £37.50

Cut and Blow Dry
"Hairy

Cost: £20.00

Discount: 35%

Saving = £7.00

Sale price = £13.00

Total Patsy Spent = £877.50

Change left = £122.50

Page 61

20 marks out of 50 in a test = 40%

30/50 = 60%

80/100 = 80%

25/40 = 62.5%

15/20 = 75%

30/40 = 75%

16/20 = 80%

Page 62/63

5 + 5 + 6 + 4 + 7 + 3 = 30 kittens

30 ÷ 6 = 5

Mean = 5

Page 64

Mean number of hours each adult cat sleeps each day = 13 hours.

Mean amount of food consumed each day by each cat family:

Maddy – 900g

Millie – 700 g

Mopsy – 500g

Moa – 700 g

Molly – 600g

Martha – 800 g

Total = 4200g, 4.2 kg

mean = 700g or 0.7 kg

It costs (42 ÷ by 100 x 40)

= £16.80 to feed the cats each day.

£16.80 x 7 = £117.60 per week.

Page 65
- Mean size of a cat family:
   6 + 6 + 8 + 5 + 7 + 4 = 36
   36 ÷ 6 = 6
- Mean number of hours each cat spends hunting for mice:
   180 ÷ 6 = 30
- Mean number of mice caught each day: 42 ÷ 6 = 7
- Mean number of hours each family sleeps: 504 ÷ 6 = 84

Page 66/67
- 30m x 10m = garden is 300m squared (300 m2)

- 2.5m x 5m = flowerbed is 12.5m squared (12.5 m2)

- 300m – 12.5m = 287.5m squared of garden left (287.5 m2)

Garden Perimeter = 80m

Flower bed = 15m.

Page 68/69

75 cm in ¾ metre, 100 cm in a metre

50 cm in ½ a metre, 25 cm in ¼ metre

Area of house = 30m2. Garage = 4.5 m2. Bin = 0.25 m2. Patio = 7m2. Pool = 12m2. Flower garden = 12.5m2. Lawn = 18m2. Shed = 4m2.

Page 70
Ingredients for 24 people
- 12.5 g sugar x 24 = 300g
- 12.5 g marg x 24 = 300g
- 12.5 g flour x 24 = 300g
- 0.25 eggs x 24 = 6 eggs

3.5 kg = 3,500g, 2.4 kg = 2,400 g

5.72 kg = 5,720 g, 4.003 kg = 4,003 g, 4.2 t = 4,200 kg, 1.1 t = 1,100 kg

6.52 t = 6,520 kg

5023 g = 5.023 kg
1203 g = 1.203 kg
72 g = 0.072 kg

Page 71
75 mm = 7.5 cm
650 mm = 65 cm
650 mm = 0.65 m
45 mm = 4.5 cm
45 mm = 0.045 m

230 m = 0.23 km

290 cm = 2.9 m

Page 72
1 kg = 1000 g
½ kg = 500 g
¼ kg = 250 g
¾ kg = 750 g

1000 ml = 1 litre
500 ml = ½ litre
250 ml = ¼ litre
750 ml = ¾ litre

1000 g = 1 kg
500 g = ½ kg
250 g = ¼ kg
750 g = ¾ kg

1000 mm = 1 m
10 mm = 1 cm
100 cm = 1 m
1000m = 1 km
1km = 1000 m

Page 73
- 320 cm
- 3,200 mm
- 3.6 m
- 3,600 mm

- 1.8 − 1.4 = Cilene jumped
0.4 m or 40cm further

Cilene ran 800 m
          80,000 cm

Kelly ran 200 m
          20,000 cm

Megan ran 40,000 cm
          400 m

Cilene ran the longest distance.
Kelly ran the shortest distance.

Page 74
3,400 g = 3.4 kg
250 g = 0.25 kg
4 g = 0.004 kg
40 g = 0.04 kg
1124 g = 1.124 kg
3204 g = 3.204 kg
506 g = 0.506 kg
70 g = 0.07 kg
1,600 g = 1.6 kg
9,200 g − 9.2 kg
1,200 g = 1.2 kg
2,400 g = 2.4 kg
3,100 g = 3.1 kg
7,240 g = 7.24 kg
3,020 g = 3.02 kg
5,001g = 5,001 kg
720 g = 0.72 kg
10 g = 0.010 kg

Page 75
What is the missing angle?
180 − (90 + 57) = 33°
180 − (70 + 35) = 75°
360 − (50 + 50) = 260° ÷ 2 = 130°

Page 76
Right Angle is 90°
Angles in triangle add up to 180°
Angle a is 180 − (90 + 42) = 48°
A straight line measures 180°
Angle b measures 180 − 42 = 138°
a = 40°, b = 70°

Page 78
Square = 4 sides
Pentagon = 5 sides
Triangle = 3 sides
Angle at centre of circle = 360°
Hexagon = 6 sides
Octagon = 8 sides
Rhombus = 2 pairs of equal angles

Page 80
25, 50, 75, 100, 125, 150, 175, 200, 225, 250, 275, 300

250, 500, 750, 1000, 1250, 1500, 1750, 2000, 2250, 2500

60, 120, 180, 240, 300, 360, 420, 480, 540, 600

90, 180, 270, 360, 450, 540, 630, 720, 810, 900

½, **1,** 1 ½, **2,** 2 ½, **3,** 3 ½, **4,** 4 ½, 5, 5 ½, 6, 6 ½, 7, 7 ½, 8, 8 ½, 9, 9½, 10

¼, ½, ¾, 1, 1 ¼, 1 ½, 1 ¾, 2, 2 ¼, 2 ½, 2 ¾, 3, 3 ¼, 3 ½, 3 ¾, 4

Page 81
3.8 + 8.17 = 11.97
7.1 + 5.62 = 12.72
13.06 + 3.5 = 16.56

5.2 + 3.42 = 8.62
15.4 + 8.45 = 23.85

34 x 5 = 170
(30 x 5 = 150)
(4 x 5 = 20)
(150 + 20 = 170)

63 x 3 = 189
41 x 4 = 164
46 x 6 = 276
76 x 2 = 152

Page 82
87 and 68 have a difference of 19

- Anna £75.00
- Sophie £50.00
- Tim £25.00

- Peter £28.00
- Kelly £14.00
- Rachel £7.00

Page 83
−2 to 7° = 9°
−3 to 4° = 7°
−1 to 2° = 3°
3 to 12° = 9°
−3 to 5° = 8°
−8 to 11° = 19°
−14 to 8° = 22°

5 to −3° = 8°
7 to −1° = 8°
3 to −2° = 5°
5 to −1° = 6°
1 to −1° = 2°

Page 84
- 60 x 50 = 3000
- 50 x 80 = 4000
- 60 x 70 = 4200

Page 85
a) 19:15
b) 23:20
c) 09:30
d) 03:15

a) 3.00 pm
b) 6.30 pm
c) 7.20 am
d) 10.15 am

Page 86

$219 \div 5 = 43\ r4$

$0.7 + 0.3 + 21 = 22$

$\frac{3}{4} = 0.75$

$4.3 = 4\ 3/10$

add 99 to 345 = 444

-5 to 4° = 9°

A factor of 54 = 9 for example.

$4/5 = 8/10 = 80/100 = 80\%$

$2/5 = 4/10 = 40/100 = 40\%$

6 ½ hours = 390 minutes

The product of 9 and 8 = 72

0.2 litres = 200 millilitres

1kg = 1000g

$80/100 = 8/10 = 0.8,\ 8/100 = 0.08$

The sum of 32 and 16 = 48

35/50 in a test = 70%

4/5 is bigger

The difference between 55 and 39 = 16

$8 \times 8 = 64$

$7\ 1/4 = 7.25$

3 less than 19.62 = 16.62

$16/5 = 3\ 1/5$

The odd number is 47

$38 \div 10 = 3.8$

The value of 6 = 600

$0.34 \times 10 = 3.4$

$4\ 1/3 = 13/3$

10 less than 16.5 = 6.5

A multiple of 7 = 42 for example.

10 more than 7296 = 7306

4/12 is equivalent to 1/3

76 rounded to the nearest 10 = 80

252 rounded to the nearest 100 = 300

The value of 2 = 20

10 more than 5092 = 5102

$0.48 \times 10 = 4.8$

Triangular numbers: 1, 3, 6, 10, 15, 21

1 m = 100 cm

Page 87

1000 years in a millennium

100 years in a century

10 years in a decade

365 days in a year

366 days in a leap year

The next leap year is in 2012 (example)

12 months in a year

52 weeks in a year

7 days in a week

14 days in a fortnight

30 days hath September, April, June and November. All the rest have 31 except February which has 28 days clear but 29 in each leap year.

Today's date = (example)

4th August 2013

4:08:2013

In 2 weeks time the date will be (example)

18th August

We are in the 21st century

Page 88

1 minute = 60 seconds

1 hour = 60 minutes

1 day = 24 hours

90° in a right angle

180° in a straight line

270° in ¾ of a circle

360° in a circle

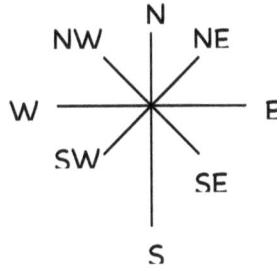

1000 grams = 1 kilogram

500 grams = ½ kilogram

250 grams = ¼ kilogram

750 grams = ¾ kilogram

Test Yourself

1) 1,000,000

2) 159 + 675 = 834

3) 908 − 66 = 842

4) 104 × 8 = 832

5) 260 ÷ 4 = 65

6) 1 ½ kg = 1500 g

7) 8,026

8) 67 p contains 50 p, 10p, 5p, 2p

9) 16:25 = 4.25 pm

10) 4.45 pm = 16:45

11) A quadrilateral has 4 sides

12) 22 cm left over

13) 0.5 = 1/2

14) 0.25 = ¼

15) 0.75 = ¾

16) 0.1 = 1/10

17) 0.01 = 1/100

18) ¾ = 0.75

19) ½ = 0.50

20) ¼ = 0.25

21) 2/10 = 0.2

22) 186 + 0 = 186

  186 − 0 = 186

23) A pentagon has 5 sides

  A hexagon has 6 sides

  An octagon has 8 sides

24) 1.7 = 1 7/10

25) 10%

26) 20%

27) 7 cm

28) 42 m − 8 m = 34 m squared

29) (6 × 7) + 14 = 56

30) 4 4/10 or 4.4

31) 60 seconds in 1 minute

  120 seconds in 2 minutes

  180 seconds in 3 minutes

32) 200

33) 11, 22, 33, 44, 55, 66, 77

  Rule = add 11 each time

34) 1, 4, 9, 16, 25, 36, 49, 64, 81, 100

35) 3.6 × 10 = 36

  3.4 × 100 = 340

36) 60 − (7 × 8) = 4

37) (36 ÷ 6) + 8 = 14

38) (9 × 7) − 10 = 53

39) 70 − (21 ÷ 3) = 63

40) 30 − (8 × 3) = 6

41) 3.58 (3u, 5t, 8h)

  14.25 (14u, 2t, 5h)

  9.234 (9u, 2t, 3h, 4th)

  18.05 (18u, 0t, 5h)

  31.624 (31u, 6t, 2h, 4th)

42) 0.10, 0.11, 0.12, 0.13, 0.14, 0.15, 0.16, 0.17, 0.18, 0.19, 0.2

3.01, 3.02, 3.03, 3.04, 3.05, 3.06, 3.07, 3.08, 3.09, 3.10

3.1, 3.11, 3.12, 3.13, 3.14, 3.15, 3.16, 3.17, 3.18, 3.19, 3.2

43)

4.5 × 10 = 45

13.8 ÷ 10 = 1.38

5 × 31 × 10 = 1,550

3.6 × 10 = 36

31.24 ÷ 10 = 3.124

71.5 ÷ 10 = 7.15

72 ÷ 10 = 7.2

9.6 ÷ 10 = 0.96

0.7 ÷ 10 = 0.07

44)

3/12 = ¼

6/12 = ½

6/18 = 1/3

1/3 = 5/15

15/20 = ¾

9/18 = ½

45) 1, 3, 6, 10, 15, 21, 28, 36, 45, 55

46) 0.5 = ½

  0.25 = ¼

  0.75 = ¾

  0.1 = 1/10

  0.01 = 1/100

47) 39 × 74 = 2,886

  82 × 46 = 3,772

  436 ÷ 3 = 145 r1

  387 ÷ 4 = 96 r3

  808 ÷ 9 = 89 r7

  529 ÷ 4 = 132 r1

  488 ÷ 3 = 162 r2

48) 4.36 x 100 = 436
    9.6 x 100 = 960
    2.31 x 100 = 231
    0.072 x 100 = 7.2
    19.1 ÷ 100 = 0.191
    80 ÷ 100 = 0.80
    71.5 ÷ 100 = 0.715
    8.2 ÷ 100 = 0.082

49) 2.05, 2.65, 5.03, 5.35, 5.7,
18.46, 19.29, 19.5, 20.05, 21.6

50) -16, -11, -6, -1, 4 = -4
   -7, 0, 7, 14, 21, 28 = +7
(rule +5)

Test Yourself 2
1) 6 x 7 = 42
6 x 8 = 48
9 x 8 = 72
7 x 7 = 49
6 x 6 = 36
7 x 6 = 42
9 x 6 = 54
7 x 9 = 63
6 x 9 = 54
8 x 8 = 64
8 x 7 = 56
7 x 8 = 56
8 x 6 = 48

2) 12, 24, 36, 48, 60, 72, 84, 96,
108, 120, 132

3) 0.5 = ½
4) 0.25 = 1/4
5) 0.75 = ¾
6) 0.01 = 1/100
7) 0.2 = 2/10
8) 0.7 = 7/10
9) 0.46 = 46/100
10) **36** – 3, 4, 2, 1, 12, 36, 18, 9, 6
   24 – 8, 6, 12, 1, 3, 4, 2, 24
   23 – 23, 1
11) 2, 3, 5, 7, 11, 13, 17, 19, 23, 29
12) 500,000
13) 250,000
14) 750,000
15) A five sided shape is a pentagon
A six sided shape is a hexagon
A seven sided shape is a heptagon
An eight sided shape is an octagon

16) 42 cm squared
17) cylinder
18) 6cm³
19) 0.2
20) 0.4

21) 0.6
22) 0.8
23) 1, 4, 9, 16, 25, 36, 49, 64, 81,
100
24) 1, 3, 6, 10, 15, 21, 28, 36, 45,
55
3/10 = 30/100 = 30% = 0.3
6/10 = 60/100 = 60% = 0.6
4/10 = 40/100 = 40% = 0.4
25) 30% 0.3, 60% 0.6, 40% 0.4
26) 11, 22, 33, 44, 55, 66, 77, 88,
27) 12, 24, 36, 48, 60, 72, 84, 96
28) ½ = 0.5
29) ¼ = 0.25
30) 1/3 = 0.33
31) 1/100 = 0.01
32) ¾ = 0.75
33) 1/10 = 0.1
34) 7/10 = 0.7
35) 1 kilogram = 1000 grams
36) ½ kg = 500 g
37) ¼ kg = 250 g
38) ¾ kg = 750 g
39) 500 ml = ½ litre
40) 750 ml = ¾ litre
41) 250 ml = ¼ litre
42) 100 cm = 1 m
43) 1000m = 1 kilometre
44) 1,000,000
45) 7 days in a week
   12 months in a year
   24 hours in a day
   48 hours in 2 days
46) 365 days in a year
   366 days in a leap year
47) AB = diameter
   CD = radius
48) angles of a triangle add up to
180°
49) straight line = 180°
50) 176 x 0 = 0
   149 x 1 = 149
   136 ÷ 1 = 136
   124 ÷ 0 = 0
51) 10 + 7 = 17
   14 + 9 = 23
   20 – 8 = 12
   20 – 4 = 16
52) 16:00 hours = 4.00 pm
53) 19 – 10p coins
54) a = 17
55) 11.45 am
56) 81
57) 300
58) mean speed was 50 miles
59) 40% of 50 = 20
60) 40 weeks
61) £105.41
62) 36 + 64 = 100
63) 0.3 > 0.03

Sally Ann Jones, wife of Peter and mother of four children trained as a teacher in the 70's and has since worked as a primary school teacher and private tutor as well as a freelance artist and illustrator. She exhibits her paintings and has published several as greetings cards.

This book is part of a series of educational material she has written based on the needs of the children she tutors. All her pupils are familiar with her cartoon drawings, which she uses as an amusing but effective way of reinforcing some basic skills.

She dedicates the book to all the children who have tested the book with such success.

www.ingramcontent.com/pod-product-compliance
Lightning Source LLC
Chambersburg PA
CBHW082307210326
41598CB00028B/4464